AGRICULTURAL PRACTICES FOR SUSTAINING MANKIND

About the Author

Professor Bimalendu Bhattacharya, completed his BA and Master's in Geography from Presidency College and the University of Calcutta, before conducting his research at the London School of Economics as a Commonwealth Scholar. He has consistently pursued a career in teaching and research, having taught at many of India's premier institutions like Gauhati University (Assam), Indian Institute of Technology (Kharagpur), Presidency College, Calcutta (now University) and the University of North Bengal where he was instrumental in founding India's first Department of Geography and Applied Geography, was a Senior Professor, Head of Department and Dean of the Faculty of Science. Prof Bhattacharya is an established Geography scholar and his publications include *Samaj Bigyaniyo Bhugol* (Social Geography in Bengali) co-authored with his late wife, Prof Anima Bhattacharya. He is the author of *Urban Development in India Since Prehistoric Times* (1979; Revised Edn. 2006), *Sikkim: Land and People* (1997), *Geography of a Himalayan Kingdom* (with Anima Bhattacharya and Neil Fraser, 2001), *Urbanisation, Urban Sustainability and the Future of Cities* (2010), *Geography of Deprivation: An Unfair World* (2014), *Global Urban Process: Growth of Towns and Cities* (2017), *Food: The Growing Challenge for Human Survival* (2020). His book, entitled *Migration Reshaping Human History* is in press.

AGRICULTURAL PRACTICES FOR SUSTAINING MANKIND

Bimalendu Bhattacharya

CONCEPT PUBLISHING COMPANY PVT. LTD.
NEW DELHI - 110 059

ISBN: 978-93-5594-171-8

First Published 2023

Published and Printed by

Concept
Since 1974

Concept Publishing Company Pvt. Ltd.
A/15-16, Commercial Block, Mohan Garden
New Delhi-110059 (India)
T: +91-11-41101460
E: publishing@conceptpub.com, W: www.conceptpub.com

Cataloging in Publication--Courtesy: D.K. Agencies (P) Ltd. <docinfo@dkagencies.com>

Bhattacharya, Bimalendu, 1931- **author.**
Agricultural practices for sustaining mankind / Bimalendu Bhattacharya.
pages cm
Includes bibliographical references and index.
ISBN 9789355941718

1. Sustainable agriculture--India. 2. Sustainable agriculture. 3. Agricultural systems--India. 4. Agricultural systems. 5. Natural resources--India. 6. Natural resources. 7. Sustainable development--India. 8. Sustainable development. I. Title.

LCC S471.I4B43 2023 | DDC 338.10954 23

Dedication

To Rupsha, Christoph and Louie
and the small farmers in the world
who will sustain their future.

Contents

Acknowledgements

I am indebted to my son-in-law, Neil Fraser, an Oxford scholar, for bringing my work to its present shape. My innuermable conversations with Neil with his long experience as a social scientist at the University of Edinburgh, has helped and encouraged me to complete this book which Neil has meticulously edited. I have also benefited from my daughter, Professor (Dr.) Bashabi Fraser, whose different background, specialising in English Literaure, has helped me to keep in view the human interest in a study of ecological background while assessing the role of small farmers as providers of food for the burgeoning global population now and in the future. My scientist granddaughter, Dr. Rupsha Fraser's critical assessment of my views have also helped me to believe in the validity and significance of the subject of this book, namely sustainable practices in agriculture that can sustain human life on this planet.

I owe a huge debt to the work of great scholars in the field who have written on various aspects of the issues, providing invaluable insights into the vulnerability of the developing world on the one hand, and its strength on the other, which I have analysed in this study. As such, I am confident that my great grandson, Louie Fraser Prossler, who holds in his little warm person the promise of the continuity of the generations, and has been an inspiring source of energy, leading me to believe that an assured future for humanity is possible through the work of small farmers who can feed humanity and ensure a future for Louie and his generation and those who will follow.

The book has been scrupulously copy edited and formatted by Dr. Saptarshi Mallick with this characteristic sincerity, for which I am truly grateful.

I am extremely grateful to Jayne Richards for her deep interest in the theme of this book and for taking on the responsibility of typing up my handwritten manuscript and make it ready for the publisher.

Bimalendu Bhattacharya

List of Tables

1
Introduction

Background of Agricultural Revolution

The Agricultural Revolution is the most remarkable one in the entire human history, much more important than any other achievement. Human life is believed to have existed for around 2 million years, but the practice of agriculture hardly goes back some 10,000 years or so and its development is not confined to one particular location but is believed to have several locations as indicated by archaeological evidence. However, South-western Asia representing most of the Middle East is taken to be the cradle of earliest agricultural practices which was the homeland of wheat and barley where the plough was invented. Similarly, besides the two crops mentioned above, the Middle Eastern area furnished rye, peas, lentils, broad beans, onions, garlic, carrots, grapes, olives, dates, apples, pears, cherries, flax and a great many other plants.[1]

Agricultural implements dating back from 9500 B.C. have been unearthed at Jericho, the Jordan Valley and in all likelihood agricultural practices extended from Asia Minor to Western Iran, Jordan, Turkey, and Cyprus. Basic ideas about agriculture further spread to Egypt, perhaps, a little before 4500 B.C. and plant domestication spread into Europe prior to 6000 B.C. with the introduction of einkorn and barley as the major crops, spreading up the Danube Valley down to Ukraine. By this time, it was also replacing foraging in the Rhineland, Belgium, Holland and Poland, reaching Denmark between 4000 and 3000 B.C., and around 3000 B.C. it had spread into most of Spain, Portugal, France, Italy including England, covering most of Europe by 2500 B.C. leaving only the coldest parts of Europe.

Agricultural practices spread in about 3500 B.C. from Iraq and Iran into South Asia and by 2500 B.C. crops like bread wheat, barley, dates and perhaps rice were grown by the flourishing civilisation of the Indus Valley called the Harappan Civilisation. Its collapse coincided with the arrival of Indo-European or Aryan-speaking barbarians from the North-west and led to the collapse of the irrigation system maintained for nearly a thousand years by the Harappans which slowed down the spread of agriculture further in the Indian Subcontinent. However, wheat was the main crop followed by millet and rice. Among other crops were oilseeds, bananas, mangoes, coconuts, and various spices, particularly in Southern India from where much later European merchants started taking spices as the major tradeable commodities.

Agricultural activities spread by way of Iran into Central Asia led by wheat and barley as major crops accompanied by millet, rice, grapes, peaches and apricots and different types of melons. According to archaeologists, agricultural practices followed the same route to reach North China.[2] Similarly, the 'overland diffusion of the South-west Asian wheat culture' benefited East Asia developing a unique crop-mix comprising 'root and rhizome (rootstock)' crops like yams, taro, and bamboo along with bananas, lotus, local millets, and sorghums, and soyabeans and rice, native to the sub-tropical Far Eastern Region. Irrigation was practised 'by about 1500-1050 B.C.', but ploughing did not begin some time before 202-220 B.C. The Central Asian trade route helped in acquiring wine-grapes, apricots, peaches and some other plants from Persia.

> Buddhist monks in China began to drink tea around 300-500 A.D. and in the course of several centuries tea-growing increased in South-east China and was introduced to Japan in the 12th century. Chinese tea was traded into Central Asia, Siberia and eventually Russia by caravan routes.[3]

Agriculture afterwards spread from China to Korea during the first millennium B.C., wherefrom the practice of rice culture entered into Japan. China further became the source of agriculture in Vietnam and partly in other parts of South-east Asia while this was also influenced by methods practised in India through trade-routes, while wet rice farming in the river valleys, coastal plains and deltaic areas seems to have started around 500 B.C.[4]

In the new world, Southern Mexico is considered to be the first centre of the agricultural revolution. Peppers and avocadoes were the first crops while in the north-east bottle gourds, peppers and pumpkin seeds were the earliest crops. The appearance of edible maize was slow to enter into the agricultural systems of what is known today as the United States in about 2000 B.C.; squash appeared in Mexico in 5000 B.C. and beans in 3000 B.C. Small scale irrigation was practised in some areas in around 900 B.C., while manioc and some types of potatoes were introduced from South America. Among other crops cultivated in Mexico were grain amaranthus, cotton, agave, cactuses, tobacco and cacao.

Plant domestication started in South America later than Mexico but in the coastal villages agriculture started being practised from 4700-3000 B.C. producing chili peppers, squash, beans and various types of root and tuber crops along with cotton, the latter earlier than in Mexico, while maize entered into the system in about 1500 B.C. On the other hand, peanut, with its origin in Brazil, spread across the Andes to the Pacific Coast and by 800 B.C. intensive farming had started in the lower coastal valleys producing new types of maize avocadoes, sweet manioc, kidney beans. By 200 A.D. canal irrigation has started in the Peruvian Coast. The Andean Valleys by this time are reported to be greatly dependant on potato and quinoa as sources of food and by the 16th century when Peru was invaded by Spanish army, Peru was producing more than 100 varieties of potato. Its origin, as claimed, goes much further back than the start of the agricultural revolution. It is interesting to know that:

> In the high valleys of the Andes, hunter-gatherers may have unearthed and consumed one such plant, potato, by 8000 B.C. but millennia passed before Indians first cultivated the tuber.[5]

Another important example of cultural expansion is provided by the extension of maize farming into Columbia by about 3000 B.C. and most of these crops were also cultivated in Chile with potato as the major crop along with sweet potatoes and sweet manioc originating in the east-Andean slopes, the production of which is presumed to go back to 1000 B.C. along with other crops like peanuts, gourds, peppers, pineapples, beans and native tobacco, dyestuffs and cotton.[6]

Overseas explorations beginning after the time of Columbus, introduced swift changes in the adoption of agricultural methods, interchange of crops, plants and farm techniques, enhanced the processes facilitated through European colonisation resulting in great transformation of agricultural processes as well as products. This is very evident in:

> maize, potatoes, better species of beans, tobacco, chilli peppers, pumpkins and squashes, new cotton strains, tomatoes, and manioc were some of the crops brought from the New World in exchange for sugarcane, bananas, yams, okra, wheat,

> oats, barley, alfalfa, wine grapes, citrus fruits, and other old world tropical or Asian crops ... while potatoes were brought from Western South America to Europe about 1565 and said to have reached Ireland by 1580. They were not grown in New England, however, until the early 18th century. By the 19th century, potatoes were becoming the mainstay of north and central European peasants.[7]

In short, global agriculture went through a revolutionary change, different crops taking very important roles, replacing others, and introducing changes in preferences in food and drinking habits of people. According to many sources, sugarcane plantations came to be centred in the West Indies. Among beverages coffee, tea and chocolate became much more popular as habitual drinks along with drastic changes in the geographical shift in coffee cultivation from Ethiopia and Western Arabia to Java, Malaya, and India to Brazil in the late 19th century. Cotton gradually came to overtake wool and flax as the major clothing material throughout the world with much expansion in its cultivation spreading to the USA, Egypt, India, Sudan, Central Asia and China.

Truly speaking, European colonisation greatly influenced agricultural policies of the countries occupied and administered by the colonial countries, such as Britain ruling over undivided India and many other countries. They determined the policy of agricultural activities primarily aimed at servicing their own interests. For example, while on the one hand destroying cottage-level industries mostly located in rural areas producing products enjoying global demand for their inimitable quality, on the other they turned the entire subcontinent into the major market for the sale of British products ranging from tea – produced in India but sold through auction in London, to cotton products made in Lancashire using cotton produced in India. This came to be the prevailing pattern of trade determining the agricultural policies of the colonised countries making overall changes in the agricultural pattern. India turned into a major provider of beverage crops like tea, coffee, or the production of cotton instead of food crops to meet local requirements and failures due to natural causes like failure of monsoon leading to occasional country-wide famines.

It is not that famines did not take place before colonial rule. There are quite a number of instances of famines having occurred in India prior to British rule happening during the earlier Muslim rule in the country, such as, a devastating one which took place in 1474 caused by failure of rains in two successive years. Further, there was the terrible famine of the most severe type that took place during Mughal rule in 1630-1632 devastating the Deccan and Gujarat preceded by two more severe ones having taken place in successive order in 1556-57 and 1573-74 affecting areas in the United Provinces (Uttar Pradesh) and the last one happening from 1594-98 was so severe as to turn people, as reported, in some areas into cannibals.[8]

In other words, all these happenings took place over a period of about three hundred years of Muslim rule in India when the country was mostly divided into many states or principalities rather than what it became – a united whole during the colonial rule lasting for nearly two hundred years during which frequency of famines was visibly much greater in spite of better communication and transport facilities with a greater control on administration. For instance, a terrible famine broke out in Bengal in 1770 victimising almost one-third of the population and in:

> the year 1866-67 a severe famine with a heavy toll of human lives spread all along the eastern coast from Calcutta (Kolkata) to Madras (Chennai). During the next year there were local famines in the United Provinces (Uttar Pradesh), the Punjab, and Rajputana (Rajasthan) in 1868-69, and in Northern Bihar in 1873. Then followed another terrible famine in 1876 which lasted for nearly two years and extended over a wide area in Madras (Chennai), Mysore (Karnataka), Hyderabad and Bombay (Mumbai) and the United Provinces (Uttar Pradesh) ... in subsequent years (occurred) the terrible famines of 1896-97 and 1899-1900. The famine of 1896-97 affected the United Provinces (Uttar Pradesh), Bihar (including present Jharkhand), the Central Provinces (Madhya Pradesh), Madras (Chennai) and Bombay (Mumbai), the area under acute distress measuring about 125,000 square miles with a population of thirty-four million. During 1899-1900 Bombay (Mumbai), the Central Provinces (Madhya Pradesh), Punjab, Rajputana (Rajasthan), Baroda and Central Indian principalities suffered in various degrees.[9]

The last of the severe famines occurred in Bengal in as late as 1943 killing a million and a half people, although it was caused by some emergency measures restricting movement of food from one to another region as well as hoardings made by the government on the pretext of fighting war with the Japanese advancing towards India across the Burma (Myanmar) Front.

The major focus here is on agriculture and its survival in a country experiencing its practice over a very long period, as mentioned earlier but its countrywide failure does not seem to be a common feature. According to Mike Davis (2014 Reprint):

> There is little evidence that rural India had ever experienced subsistence crises on the scale of Bengal Catastrophe of 1770 under East India Company rule or the long siege by disease and hunger between 1875 and 1920 that slowed population growth almost to a standstill … On the other hand, benefiting perhaps from under a milder ENSO (El Nino – Southern Oscillation).* Mugul India was generally free of famine until the 1770s. There is considerable evidence, moreover, that in pre-British India, before the creation of a railroad–girded national market in grain, village-level food reserves were larger, patrimonial welfare more widespread, and grain prices in surplus areas better insulated against speculation … Recent research by Ashok Desai indicates that 'the mean standard of food consumption in Akbar's empire (1560-1605) was appreciably higher than in the India of early 1960.[10]

Indeed, the British claim of rescuing India from 'timeless hunger' during its rule in India is absolutely baseless in view of the facts of contrasting situation of 31 serious famines happening in: '120 years of British rule in India against only 17 recorded in the entire previous two millenia'.[11]

As a matter of fact, it is stated by the same author basing his views on the findings of others that Mugul rulers used to take a very much benevolent view of the farmers suffering at the hands of natural calamities evoked by ENSO-related droughts and floods by way of controlling food exports, price regulation, tax relief along with free distribution of food to affected farmers without any forced labour counterpart. It was, however, in sharp contrast to the imposition of punitive taxation of irrigation and its neglect of traditional wells and reservoirs by the Raj or the colonial power in contrast to tax subsidies used by the Muguls to promote water conservation. On occasions, however, the ruling British acknowledged the benevolent attitude of the Mugul emperors at the time of the national crises resulting from natural disasters, such as the First Famine Commission Report in 1880 refers to the extraordinary relief provided by Emperor Aurangzeb at the time of drought – famine in 1661 caused by El Nino:

> The Emperor opened his treasury and granted money without stint. He gave encouragement to the importation of corn and either sold it at reduced prices or distributed it gratuitously

> amongst those who were too poor to pay. He also promptly acknowledged the necessity of remitting the rents of the cultivators and relieved them for the time being of other taxes. The vernacular chronicles of the period attribute the salvation of millions of lives and the preservation of many provinces to his strenuous exertions.[12]

According to this source, the occasional failure of the local farming system to feed the local population resulted from the farmers and peasants becoming victims of national disasters after 1850s and the forcible incorporation of their small production system along with its overall control taken over by overseas financial circuits undermining the traditional security.

But, the importance of the local farming system may be gauged from the fact that in the middle of eighteenth century India and China provided the largest manufacturing districts in the world located on the Yangzi delta and Bengal with Madras (Chennai) closely following while in South India labourers had higher earnings than their British counterparts in the eighteenth century and lives of greater financial security and South Indian Weavers and other artisans 'enjoyed better diets than average Europeans'.[13]

Such evidence undoubtedly supports the view that agricultural producers, that is the farming community, were undoubtedly much better off than they were in India during the British rule.

But worldwide food-shortage, known as the Great Hunger of 2008 with unprecedented increase in corn-prices following fall in corn harvests in the United States by fifty per cent compared with the year before, led further to the rise in its price along with prices of other commodities like soyabeans, wheat, rice and vegetable oils simultaneously raising prices of basic foodstuffs two to three times in a short period giving rise to food riots in Mexico and many other countries of the global south.

The reasons behind such catastrophic situation are not difficult to trace. Such as increased diversion of especially corn and soyabeans to meet growing meat consumption which doubled between 1960s to 2007, depeasantisation policies of some countries for controlling food prices and thus forcing peasant farmers off the land and into urban slums, increased control of corporate houses in the diversion of food crops, like corn, soyabeans, palm oil (and even wheat) into the production of ethanol to replace petroleum along with major crop failures, such as in Bangladesh in 2008 due to devastating droughts followed by Russia's annual wheat harvest affected by unprecedented heatwaves along with Pakistan's wheat and rice crops threatened by unprecedented floods in 2010 added very strongly to food-security concerns particularly for the peasantry.[14]

Another very serious side of the improvisation of the farmers worldwide affecting almost 20 to 30 million who lost their land to capitalistic export-based agriculture followed a policy of land-grab. In this connection, it is important to note that there are about '3.2 billion hectares of land or 7.86 billion acres potentially suitable for cultivation, about half of which is under cultivation'.[15] But the interest of corporate houses in their attempt to take away agricultural land from the hands of the peasant farmers in connivance with local governments and convert the land grabbed into the production of commercial crops like maize, soyabeans or palm oil often using genetically produced seeds instead of indigenous ones acclimatised through generations of use with natural elements. This led to major attempts of land-grab in cases like the South Korean Firm, Daewoo Logistics buying 1.3 million hectares of arable land in Madagascar to produce 5 million tons of corn by 2023 and palm oil from another lease of 1,20,000 hectares. Countries, besides South Korea joining in this spree of grabbing land for the production of industrial crops were China, Japan, Saudi Arabia, United Arab Emirates etc. together grabbing up 76,23,673 hectares and major countries involved in giving up farmers' land were mainly Sudan, Indonesia, Mongolia, Philippines, Laos, Brazil, Pakistan etc. The entire process was termed by the Head of the FAO creating 'neocolonialism'.[16]

Interestingly, land-grabbing has been going on for centuries starting with the discovery of America and brutal expulsion of the indigenous communities from their ancestral properties by the white colonisers, the same process repeated later time and again by colonial powers on different pleas or excuses.

> As of early 2010, well over 40 million hectares have exchanged hands or are under negotiation – 20 million of which are in Africa. GRAIN calculates that over $100 billion have been put on the table to make it happen. Despite the governmental grease here and there, these deals are mainly signed and carried out by private corporations in collusion with host country officials.[17]

Investing in agriculture is more and more becoming a big business involving tens of billions of dollars and rising rapidly. It is not turning agriculture to solve world hunger and eliminate rural poverty but changing it into agri-business operations.

> Today's global food crisis will not be solved by large-scale industrial agriculture.... This agriculture will not feed the planet: it will feed speculative profits for a few. There will be seemingly more secure food for the wealthy, and more poverty for the rest .... Never has the world's food supply been more highly controlled by a small group, whose decisions are based solely on how much money they can extract for the shareholders.[18]

Unfortunately, this is where the major part of the humanity is being led to and unless this is restrained without further delay even the rich people will feel the pinch in no time as production of crops needed for every human body will gradually become scarce for even money to buy.

Revival of the practice of agriculture means primarily for direct human-consumption which needs to be encouraged in every society as an action of love for some people who enjoy the very touch of soil and make it produce edible food crops as induced by the handler. We should put all our efforts at sustaining as well as encouraging agricultural practice by a section of people making their living on it which alone is capable of sustaining mankind faced today with chronic shortage of food more in the developing world though not less in the developed world in its affluence having greater demand for fruits and choice vegetables.

In short, production of food for the consumption of every person can no longer be shelved on false pretences when shortage of food is becoming an unavoidable consequence threatening human survival.

NOTES

***Davis, Mike**. *Late Victorian Holocausts: El Nino Famines and the Making of the Third World,Verso*. 2014: p.13. Jacob Bjerknes at UCLA in late 1960 'showed for the first time how the equatorial Pacific Oceans, acting as a planetary heat engine coupled to the trade winds, was able to affect rainfall patterns throughout the tropics and even in the temperate latitudes. Rapid warmings of the eastern tropical Pacific (called El Nino events) for example, are associated with weak monsoons and synchronomous drought throughout vast parts of Asia, Africa, and north-eastern South America. When the eastern Pacific is unusually cool, on the other hand, the pattern reverses (called as a La Nina event), and abnormal precipitation and flooding occur in the same *teleconnected* regions. The entire vast see-saw of air mass and ocean temperature, which extends into the Indian Ocean as well is formally known as *El Nino-Southern Oscillation* (or ENSO, for short)'.

1. *Encyclopaedia Americana*, New York: Americana Corporation, Vol. 1, 1973, p. 354.
2. *Ibid.,* p. 356.
3. *Ibid.*
4. *Ibid.*
5. Rhoads, Robert E. "The World's Food Supply at Risk," *The National Geographic Society,* Washington, 179.2 (1991), p. 77.
6. *Encyclopaedia Americana*, New York: Americana Corporation, Vol. 1, 1973, p. 357.
7. *Ibid.,* p. 358.
8. Majumdar, R.C., Ray Chaudhuri, H.C., Datta, K. *An Advanced History of India*. Delhi: Macmillan, 1970, pp. 364, 465-466, 564, 864.
9. *Ibid.,* pp. 864-866.
10. Davis, M. *Late Victorian Holocausts: El Nino Famines and the Making of the Third World.* London and New York: Verso, 2014, pp. 285-286.
11. *Ibid.,* p. 287.
12. *Ibid.,* p. 286.
13. *Ibid.,* p. 292.
14. Magoff, Fred, Tokar, Brian. "Agriculture and Food Crisis: An Overview" in *Agriculture and Food Crisis: Conflict, Resistance and Renewal* ed. Magoff, Fred, Tokar, Brian. *Monthly Review*, 2010, pp. 9-11.
15. Bhattacharya, Bimalendu. *Geography of Deprivation: An Unfair World.* New Delhi: Concept Publishing Company Pvt. Ltd., 2014, p. 23.
16. *Ibid.,* pp. 23-24, 25.
17. GRAIN. The New Farm Owners: Corporate Investors and Control Overseas Farmland. 2010 in Magoff, Fred, Tokar, Brian. *Agriculture and Food Crisis: Conflict, Resistance and Renewal.* Monthly Review, New York, 2010, pp. 140-141.
18. *Ibid.,* p. 147.

2
Natural Wealth
Possibilities for Strengthening the base of Farming Practices

I. The Role of Natural Wealth

In order to assess the possibility of taking up the challenge of turning to agricultural practices as the only way to save humanity from the impending mass-food crisis it becomes imperative to take a stock of the natural capital of the world which on their own may contribute to the way of approaching the major issue of putting stress on agriculture, or rather the farming practices at the peasant-level as the determinant of human survival.

Anything that nature offers – be it soil, water, climate, vegetation, minerals, or even the natural sites like a plain land, river valleys, hills and mountains, a sea beach, a waterfall or spring, the animals and the lowliest of insects, its biodiversity consisting of a world of natural elements, to name a few – has some use for human life in deriving some benefit for us in prolonging our existence or, rather human life and leave enough for use of our future generations to make a life of their own[1].*

However, today's consumerist, corporate-centric industrialisation of crop-economy has so far failed to take the severe warning of leading human-food position to virtual collapse (not very far away), ignoring the caution sounded not very long ago by the team of experts in their 'Report for the Club of Rome' some time in 1972 focusing on 'The Predicament of Mankind'.

> The intent of the project is to examine the complex of problems troubling men of all nations: poverty in the midst of plenty; degradation of the environment; loss of faith in institutions; uncontrolled urban spread; alienation of youth; rejection of traditional values; and inflation and monetary and economic disruptions … the team examined the five basic factors that

> determine and therefore, ultimately limit the growth of this planet – population, agricultural production, natural resources, industrial production, and pollution.[2]

That the supply of natural resources is not limitless but man in his lust for making life more enjoyable has crossed all limits failing to realise that the highly insensible rate of use will lead to severe crises in no time as we feel the major pinch in the fast rate of climate change almost upsetting all natural processes involving human life. Ultimately devastating floods, unprecedented droughts, melting of sea ice, rising sea-level jeopardising coastal areas, thawing of permafrost areas, destruction of mangrove forests, severe dust storms, rise in temperature and fast depletion of groundwater resources in many areas in response to rising demand resulting from lack of precipitation are only a very few of the manifestations of climate change hitting very hard on life-supporting activities.

Fast depletion of natural resources in the urge for 'present' comfort has overlooked the consequences. Interestingly, the very title of the 'Report', 'Limits to Growth' carried an inherent veiled threat made by the authors of the 'Club of Rome', preparing this masterpiece challenging reckless handling of nature. The major problem raised in this 'Report' was 'inadequate nourishment' of almost 60 per cent of the population living in less industrialised countries with deficiency in basic caloric requirements. According to this study, there were about '3.2 billion hectares (7.86 billion acres) of potentially suitable land for agriculture' about half of which – the 'richest, most accessible' – was under cultivation, the remaining land requiring 'immense capital-inputs to clear, irrigate and fertilise' before put to agricultural production.[3]

As estimated in this 'Report', if people decided to cultivate all possible arable land at that time (around 1972) to feed, 'the present world average of 0.4 acres per person is sufficient (to feed the entire world population at present US standards 0.9 hectares per person would be required).[4]

The same 'Report', however, cautioned that each additional person requires 0.08 hectares per head to cover all requirements besides land producing food, for such uses like 'housing, roads, waste disposal, power lines and other essential uses. This, however, presumed that 'advances in agricultural technology in future' might increase yield rate of food crops twice or even four times with larger investments in 'tractors, fertilisers and irrigation' systems. It is interesting to know:

> To achieve a 34 per cent increase in world food production from 1951 to 1966, agriculturalists increased yearly expenditure on tractors by 63 per cent, annual investment in

> nitrate fertilisers by 146 per cent, and annual use of pesticides by 300 per cent. The next 34 per cent increase will require even greater inputs of capital and resources.... The supply of food to be expected in the future is dependent on land and fresh water and also on agricultural capital.... Opening new land, farming sea, or expanding the use of fertilisers and pesticides will require an increase of capital stock devoted to food production. The resources that permit growth of that capital stock tend not to be renewable resources, like land and water, but non-renewable resources like fuels and metals. Thus, the expansion of food production in the future is very much dependent on the availability of non-renewable resources.[5]

It is important to note in this context that we have to be extra-careful from now on in exercising maximum restraint on further exploitation of our environment and further impair the life-supporting capacity of the earth. Such a restraint is likely to provide 'reasonably large populations with a good material life plus opportunities for limitless individual and social development'.[6]

Dwelling so far on the limitations of the Earth's support system to human life, or rather not to jeopardise its continuation in the face of man's inexhaustible pressure on its limited resources along with his more exploitive attitude, one is surprised as well as appalled to notice that in the name of an 'advanced economy' man has taken an absolutely wrong path leading to 'rather' destruction of his own achievements. Donella Meadows, the lead author of *The Club of Rome* has expressed this in no uncertain terms as a last try (dying in 2001) to save humanity from its greatest delusion in the following words 'as the last warning that Earth's resources are not inexhaustible:

> The first commandment of the earth is not *Enough.* Just so much, and no more. Just so much soil. Just so much water. Just so much sunshine. Everything born of the earth grows to its appropriate size and then stops. Its creatures learn, mature, diversify, evolve, create amazing beauty and novelty and complexity, but live within *absolute limits'*.[7]

The message about the 'absolute limits' is very clear. Hence, man has to understand that under no condition the limit of nature's bounty should be crossed in the name of so called 'development', taking for granted that science cannot replace nature and create what is lost from Nature. In this perspective it now becomes obligatory to take a stock of 'Natural Capital'.

Natural capitals of various types may be considered as assets providing goods and services for ultimate benefit of humans and they may be non-renewable such as oil, gas, minerals etc. or renewable such as the habitat of a particular bird, say Nightingales. The first thing in assessing the natural capital of a country is its population structure which, according to Dieter Helm (2015) will be very different by 2050 compared to now, since people in general are living longer consisting of more centenarians 'consuming and not producing' and making greater demands for services. On the other hand, global economic growth being about 3-4 per cent per annum now 'doubles every 15-20 years, and granting consumption remains the same the world 'will be consuming twice by 2030' than what it is doing now.[8]

At the same time, climate change and its impacts on agriculture will be staggering making it harder to feed 9 or 10 billion people requiring revolutionary changes in agricultural practices keeping in step with growing population as well as growing demand with double the requirement for meat, fish and nutritional food spurred further by growing income in many developing countries. It is observed that agriculture will take an important place in meeting the growing requirement of animal protein, but such a process is going to be largely hampered by destruction of mangroves in South-East Asia for shrimp farming. Another problem is faced by the use of growing proportion of food crops like corn, sugarcane, and palm oil and even wheat – as in Britain – for the production of ethanol, already using 40 per cent of corn in the USA, as mentioned earlier. Practices like this 'not only push up food and land prices along with changes in land-use pattern, but also affects the pattern of carbon emissions. On the other hand, 'trying to double food production' will severely affect biodiversity. The extent of the damage may be ascertained from the fact that '25 biodiversity hotspots' of the world originally covering about 12 per cent of the global land surface has got reduced to only '1.4 per cent' which has further lost much of the primary vegetation along with severe decline in the population of mammals, birds, reptiles, amphibians, fish, 'having declined on an average of 52 per cent since 1970'.[9]

As the existing situation indicates, further destruction of the natural environment has to be stopped to avoid further interference with natural resources getting fast devoured by what we call economic development proving again the dire warnings of Malthus made about 200 years ago. Helm (2015), however, is a bit sceptical about 'The Club of Rome's observations about the imminent dangers of the environment through its reckless use by mankind. He strongly believes that we are not yet as endangered as warned by 'The Club of Rome'. In other words, Helm

(2015) believes that we can yet forestall the situation with means already developed by man. Helm (2015) accepts that:

> The damage (obviously to environmental or existing resources) done so far is great, and much greater damage is on its way in the business-as-usual scenario, and it can and will go on for quite some time to come unless we change the current economic path. This means treating the environment not simply as a constraint on economic activity, but rather as an integral part of the economy. The starting point of such an exercise ... is to work out what a sustainable growth path looks like, and the part that natural capital plays in underpinning it.[10]

As one of the ways for achieving such a position, as mentioned above, without seriously interfering with the environment or further depleting natural resources or capitals, Helm (2015) points out that change in food-habits, such as adopting vegetarianism widely as a food-habit will cut down consumption of crops required to rear animals, thus making a direct check on their widespread production and reduce corresponding emission of CO_2 into the atmosphere thereby making a lasting effect on increase of temperature as a source of climate change; or 'compulsory recycling of water and materials besides incorporating change in irrigation practices replaced by the introduction of drip irrigation to save water as well as no-till farming practices with drastic reduction in the use of chemical fertilisers and harmful insecticides along with limited travel, especially aviation, relying more on walking, holidaying at home rather than abroad would further reduce the environmental-footprint'.[11]

The concept of sustainable development, perhaps, depends very largely on a judicious way of utilising the environment for economic development leaving enough for the future generations, as advocated by the Brundtland Commission Report (1987) so that they do not become handicapped in making economic progress. It, however, emerges to be a very complicated situation in estimating how much environmental capital, say the forests, to be used and how much to be left undisturbed. Of course, forests once used up can be renewed through afforestation in a meticulously calculated manner but does not apply to non-renewable assets or natural capitals like fossil fuels or minerals which have been created by nature through a complicated process over millions of years. This equally applies to another very important natural resource or capital, which is underground water already depleted mercilessly, as mentioned in the previous chapter (Introduction) without any care for even the present generation. At the same time, the argument against the warning of '*The Club of Rome*' that it did not foresee the progress of science to combat the theory of overuse of natural assets giving rise to environmental deficiency does not sound tenable. Scientific achievements cannot solve the problem of soils and water getting fast depleted and thus pushing the entire humanity to dire consequences in the near future when both are far from adequate to grow food for the mankind unless the rate of exhaustion checked immediately. However, science could do something very positive about further growth of global population devising useful means and their application in the scientific manner in the right direction without coercion and prevailing upon people with proper reasoning. Such as, use of contraceptives made simpler and more encouraging with proper guidance, universal education with particular emphasis on females, discouraging early marriage among girls, and easy provision of health and hygiene particularly aimed at developing countries. Nobody can assure that science cannot make the situation capable of producing adequate food for the entire human community with its growing demand without making drastic changes in its preferences putting greater stresses on stemming the sources of carbon emissions lying at the root of rise in temperature causing climate change. One of the effective ways of dealing with both the problems reducing the release of carbon dioxide emissions in the atmosphere and thereby inducing climate change is to make a drastic reduction in the use of coal, the major fossil fuel in use today worldover including the USA, China and India followed by storing the coal-gas using the CNG process to be discussed later.

According to UNDP Report 2007/2008, in order to fight climate change which is at the root of all maladies risking human life very seriously is CO_2 emissions, mainly from countries like the United States with five times carbon footprint that of China and India while it is 0.2 tonnes of CO_2 in Ethiopia compared with 20 tonnes in Canada. Thus, to keep to the threshold of 2°C temperature of pre-industrial level rich nations are required to cut emissions by at least 80 per cent, with cuts of 30 per cent by 2020 while emissions from developing countries peaking around 2020, with cuts of '20 per cent by 2050'.[12]

On assessing the records of climatic change disasters as an engine of death and destruction caused by climate change it needs to be pointed out that they are, so far, concentrated more in poor countries affecting some '262 million people annually from 2000 and 2004', comprising over 93 per cent of them in the developing countries. While frequency of them is on increase with unbridled rise in emission, four key transmission mechanisms are likely to decelerate the rate of climate change. They are:

(a) Agricultural Production and Food Supply

Climate change affects rainfall, temperature, and water availability for agriculture drastically expanding drought-affected areas, as in Sub-Saharan Africa, or even in major parts of Latin America and South Asia with crop-failures and thus increase the volume of undernourished people already plagued by associated problems of health and hygiene likely to add 600 million more by 2080 to rural poverty.

(b) Water Stress and Water Insecurity

Increasing rate of glacial melt caused by rising temperature already causing retreat by 10-15 metres a year in the Himalayas have a catastrophic effect on seven of Asia's great river systems with similar effect in the Andean region as well as in the Middle East facing 'deep losses in water availability'.

(c) Rise in Sea Levels and Exposure to Climate Disasters

Rapid ice-sheet disintegration, as is reported to be already happening in the Antarctic with a massive chunk dislodged and slowly melting down, will cause serious change in sea level when a rise in temperature by 3-4°C caused by increase in CO_2 emissions could make 330 million people permanently dislodged, directly affecting 70 million in Bangladesh, 6 million in Lower Egypt and 22 million in Vietnam besides many more living in place like Laccadive Island close to India, and small island states in the Caribbean and Pacific await an unpredictable future already exposing over 344 million people in these regions to tropical cyclones and more intense storms like a cyclone with over 80mph wind speed at times completely devastating the lives of West Bengal.

(d) Ecosystems and Biodiversity

The impact of climate change already has devastating effects on almost one-half of the global coral reef systems with a 'bleaching' effect resulting from warming of seas while increasing acidity makes a great threat to marine ecosystems.[13] Thus, the effect of climate change on human development may be summarised in the following manner:

First, in terms of human development, it is clearly pointed out that the effect of climate change will be felt much more in developing countries, particularly in Asia, and on the whole 2.6 billion or 40 per cent of the global population earning less than US$ 2 a day will be severely affected by it.

Second, around 28 per cent of all children of the developing countries, particularly those in South Asia and Sub-Saharan Africa are 'underweight or stunted' because of malnutrition.

Third, there will be high rate of child mortality, almost 10 million dying each year before they are 5 years old resulting from poverty and malnutrition in South Asia and Sub-Saharan Africa. Only 52 out of 147 countries are expected to achieve a two-thirds reduction by 2015.

Fourth, the life of poor people across the world is subject to infectious diseases with about 40 million subjected to HIV/AIDS, 3 million of them dying in 2004. Besides, there are 350-500 million subjected to malaria with 1 million deaths every year, Africa accounting for 90 per cent of them.[14]

The effect of climate change primarily resulting from crossing the 2° C threshold is going to be catastrophic in various fields directly hitting human (in fact, all living objects) life in different ways:

(i) Agriculture will be the first victim with losses in food-supply linked to severe 'drought' and rainfall variations in parts of the Sub-Saharan Africa, South and East Asia making another 600 million people subject to acute malnutrition by 2080.

(ii) Crossing the 2°C threshold will, as mentioned above, accelerating glacial melt in the Himalayas, particularly, will create serious ecological disorders across China, India, and Pakistan, by increasing floods besides reducing the volume river-flows felt equally in Latin America cutting supplies of water for agriculture, hydroelectricity and urban use. The effect of climate change will be devastating, subjecting 1.8 billion people facing water shortage by 2080.

(iii) The third most important effect of climate change is expressed in the severity of coastal flooding and extreme weather events. An average rise in temperature in excess of 2°C threshold will give rise to violent storms causing widespread displacement of people affected mostly in coastal or low-lying areas of countries like – as already mentioned – Bangladesh, Egypt, Vietnam etc., displacing almost 262 million people between 2000-2004. The number of so many people becoming victims of the sea-level rise and tropical storm may increase by another 180 to 230 million.

(iv) It is predicted that temperature-increase exceeding the threshold limit to 3°C would result in the collapse of ecosystems causing extinction of 20-30 per cent living species, particularly resulting in the obliteration of coral reef systems, as mentioned above, through extensive 'bleaching'.

(v) Health-risk resulting from climate change, particularly increasing the risk of malaria, as mentioned above, would involve additional 200-400 million people.[15]

According to the Report of the United Nations (2003), water has already turned into a very scarce commodity, particularly when 'more than 2 billion people' suffer from shortage of water in over forty countries, 'one billion do not have sufficient drinking water' and '2.4 billion' do not have any provision for sanitation. It is necessary to note that the per capita annual requirement of drinking water for maintaining a healthy life is 1,700 cubic metres.

But due to population growth per capita water supplies 'decreased by a third between 1970 and 1990' and the situation will aggravate further with a much larger population in future, putting almost '7 billion people in sixty countries' under severe water-stress by 2050.[16]

In this regard, Lester R. Brown (2012) is much more explicit about the worsening situation pointing out that there is no substitute for water to grow food and although an adult consumes about 4 litres of it a day, it takes 500 times as much or 2000 litres to 'produce the food we consume each day'. It is for this reason that 70 per cent of world water is used for irrigation and this has been increasing further, having been very fast in the second half of the twentieth century rising from 250 million acres world-over to about 700 million acres by 2000 though slackened since then 'expanding only 9 per cent' between 2000 and 2009. The increase, however, has never been commensurate with population-growth, with a per capita decrease of 10 per cent than it was in 1960. As observed by Brown (2012), the fast depletion of many aquifers with reckless over pumping, the situation will worsen further in coming years with gradual decline in grain harvest dependent on irrigation. According to Lester R. Brown (2012), the countries already over-pumping the aquifers by 2012 are – Afghanistan, China, India, Iran, Iraq, Israel, Jordon, Lebanon, Mexico, Morocco, Pakistan, Saudi Arabia, South Korea, Spain, Syria, Tunisia, United States and Yemen. It is interesting to observe that of the 40 per cent of world grain grown by irrigation, the proportion varies among the three major producers, such as it is four-fifths of the grain harvest in China, three-fifths in India and only fifth in USA,[17] the remaining crops grown in each of them being rain-fed.

On the other hand, according to this same source the spree for dam-building on rivers channelling water for irrigation having diminished in the 1970s, attention was diverted to drilling wells for tapping underground water sources for irrigation, which depleted the resources in the long run where use exceeded the rate of replenishment. Countries like Saudi Arabia, Ethiopia, Sudan, Yemen, Jordan, China, and India provide such examples of overpumping and among them India makes a special case on two counts, such as while rich farmers use 'modified oil-drilling technology' to pump out water without any check, on the other extreme 'a thriving watertruck industry has emerged' where:

> for farmers near the city, the market price of water far exceeds the value of the crops they can produce with it. Unfortunately, the 13,000 tankers hauling water for Chennai (a city of 9 million) are mining the region's underground water resources. Water tables are falling, and shallow wells have gone dry.[18]

Since water is one of the basic requirements of agricultural production it is necessary to make an assessment of the availability of water from different sources. The most primary source of water is precipitation or rainfall, the large portion of it happening in the tropics, with annual totals of 2,400 millimetres and more in the mid-latitudes close to high mountain ranges. Such precipitation takes place in the form of monsoon, tropical cyclones, and convective storms. However, towards the poles, or with increasing altitude, precipitation takes place in the form of snow, the annual global snowfall amounting to 1.7×10^{13} tons covering an area ranging from 100 to 126 million km^2. Total annual precipitation on the land surface all over the globe is estimated to be 1,19,000 cubic kilometres (km^2) (U.N., 2003). According to the same source (Shiklomanov, 1998[a]) annual loss of moisture through evaporation from the land surface of the globe amounts to a total of about 74,000 km^3.

> Besides precipitation, another very important source of water is groundwater or huge amount of fresh water stored in subterranean reservoirs, called aquifers, exploitation of which on a massive scale is a recent development with major advances in 'geological knowledge, well drilling, pumping techniques' etc. initiating rapid exploitation between 1950 and 1975. With global exploitation at the rate of 600-700 km^3 groundwater is a very important source of irrigation in Asia, provides about 70 per cent of piped water supply in the European Union, and supports rural livelihoods across extensive areas of Sub-Saharan Africa (U.N., 2003). The groundwater makes a predominant reserve of freshwater estimated at 70,00,000 km^3 to 23.4 million km^3 providing

about 50 per cent of current potable water supplies, 40 per cent of industrial requirements and 20 per cent for irrigated agriculture. At the same time over 1.2 billion urban dwellers depend on well, borehole and spring sources (Table 2.1).

In India, ground water directly supplies about 80 per cent of domestic water supply in rural areas, with some 2.3 to 3.00 million hand-pump boreholes having been constructed over the past thirty years. Further, some 244 km^3/year are currently estimated to be pumped for irrigation and from about 15-17 million motorised dug wells and tube wells, with as much as 70 per cent of national agricultural production being supplied by groundwater.[19]

Another very important source of fresh water-reserve of the globe are the icecaps and glaciers, about 90 per cent of which is concentrated in Antarctica while most of the remainder is in Greenland. With the small icesheets, and glaciers in North and South America, Europe and Asia, the total volume of ice across the globe over an area of 16.2 million km^2 with total volume of water across the globe exceeding 24 million km^3.

Table 2.1: Some Large Aquifers of the World

No.	Name	Area (million km^2)	Volume (billion m^3)	Replen-ishment Time (Years)	Conti nent
1.	Nubian Sandstone Aquifer System (Africa)	2.0	75,000	75,000	Africa
2.	North Sahara Aquifer System (Africa)	0.78	60,000	70,000	Africa
3.	High Plains Aquifer (North America)	0.45	15,000	2,000	North America
4.	Guarani Aquifer System (South America)	1.2	30,000	3,000	North America
5.	North China Plain Aquifer System (Asia)	0.14	5,000	300	Asia
6.	Great Artesian Basin (Australia)	1.7	20,000	20,000	Australia

Source: The United Nations World Water Development Report, 2003, Table 4.3: p. 79.

Table 2.2: Use of Groundwater for Agricultural Irrigation

Country	Irrigated Area (M ha)	Irrigation Use (km^3/year)	% of Groundwater
India	50.1	460	53
China	48.0	408	18
Pakistan	14.3	151	34
Iran	7.3	64	50
Mexico	5.4	61	27
Bangladesh	3.8	13	69
Argentina	1.6	19	25
Morocco	1.1	10	31

Source: The United Nations World Water Development Report, 2003, Table 4.4: p. 80.

On the other hand, the permafrost 'extending over north-east Europe, northern Asia, northern Canada and the fringes of Greenland and Antarctica and the higher parts of South America' makes a total of 21m km^2 though water from this makes little contribution for economic purpose.

But according to UN (2003) Report, there were 47,665 large dams in the world in 1995 and another 8,00,000 smaller dams (less than 15 metres in height) and there are 633 large reservoirs. These reservoirs behind large dams across the globe cover a total area of about 1,68,000 km^3 of water. The reservoirs, however, have problems which does not appear on the surface but sediment accumulation in them behind the dam taking a frightening level induces great problems. Professor Khalid Mahmood of George Washington University in Washington DC, making rough estimate (1987) for World Bank Study stated that:

> around 50 cubic kilometres of sediment – nearly 1 per cent of global reservoir storage capacity – is trapped behind the world's dams every year. In total, calculated Mahmood, by 1986 around 1,100 cubic kilometres of sediment had accumulated in the world's reservoirs, consuming almost one-fifth of global storage capacity.[20]

It may, further, be interesting to know that according to Mahmood, dredging of sediments accumulated in reservoirs worldwide would come to $100-$150 billion (at the price level of his time). But much more frightening prospect resulting from the huge weight of the sediments lay in triggering earthquakes. According to his prediction:

> Today there is evidence linking earth tremors and reservoir operation for more than 70 dams. Reservoirs are believed to have induced five out of nine earthquakes on the Indian peninsula in the 1980s which were strong enough to cause damage.[20] (Table 2.3)

On the other hand, lakes numbering about some 15 million worldwide are of different origin, some contain fresh water while some have saline water. At the same time, rivers representing the most accessible and important source of fresh water 'though in many cases much of the run-off involving even some of the largest rivers of the world is lost for various reasons and as such making human-dependability quite unsteady'. For example, rivers crossing arid and semi-arid regions, such as 'Indus (Pakistan), the Niger (West Africa), the Nile (Eastern Africa), and the Colorado (Argentina), together giving rise to a total loss of 1100 km^3 run-off per year'.

The world's largest river, the Amazon in Latin America, contributes some 16 per cent of the global total run-off, while the five largest river systems (Amazon, Ganges with the Brahmaputra in India, Congo in Central Africa, Yangtse in China and Orinoco in Venezuela) together account for 27 per cent.[22]

Table 2.3: Reported cases of Reservoir-induced Seismicity Greater than Magnitude 4.0 (Richter Scale)

Dam	Country	Dam Height (m)	Reservoir Volume ($m^3 \times 10^6$)	Impounding Began	Largest Earthquake	Size
Koyna	India	103	2,780	1962	1967	6.3
Kariba	Zambia/ Zimbabwe	128	1,75,000	1958	1963	6.2
Kremasta	Greece	160	4,750	1965	1966	6.2
Xinfen-gjiang	China	105	14,000	1959	1962	6.1
Srina Kharin	Thailand	140	17,745	1977	1983	5.9
Marathon	Greece	67	41	1929	1938	5.7
Oroville	USA	236	4,400	1967	1975	5.7
Aswan	Egypt	111	1,64,000	1964	1981	5.6
Benmore	New Zealand	110	2,040	1964	1966	5.0

contd...

Dam	Country	Dam Height (m)	Reservoir Volume ($m^3 \times 10^6$)	Impounding Began	Largest Earthquake	Size
Eucumbene	Australia	116	4,761	1957	1959	5.0
Hoover	USA	221	36,703	1935	1939	5.0
Bajina-Basta	Yugoslavia	90	340	1966	1967	4.5-5.0
Bhatsa	India	88	947	1981	1983	4.9
Kerr	USA	60	1,505	1958	1971	4.9
Kurobe	Japan	186	149	1960	1961	4.9
Monteynard	France	155	275	1962	1963	4.9
Shenwo	China	50	540	1972	1974	4.8
Akosombo	Ghana	134	1,48,000	1964	1964	4.7
Canelles	Spain	150	678	1960	1962	4.7
Danjiangkou	China	97	16,000	1967	1973	4.7
Grandval	France	88	292	1959	1963	4.7
Kastraki	Greece	96	1,000	1968	1969	4.6
Lake Pukaki	New Zealand	106	9,000	1976	1978	4.6
Nurek	Tadjikistan	317	10,500	1972	1972	4.6
Fuziling	China	74	470	1954	1973	4.5
Khao Laem	Thailand	130	8,860	1984	1985	4.5
Piastra	Italy	93	13	1965	1966	4.4
Lac de Vouglans	France	130	605	1968	1971	4.4
Clark Hill	USA	60	3,517	1952	1974	4.3
P.Colombia/ Volta Grande	Brazil	40/56	1500/2300	1973/4	1974	4.2
Camarillas	Spain	49	37	1950	1964	4.1
Manicouagan	Canada	108	10,423	1975	1975	4.1

Source: Patrick McCully, *Silenced Rivers: The Ecology and Politics of Large Dams*, Orient Longman: Table 4.1: p. 125.

II. Human Role in Climate Change and Mitigating Measures

Apart from a wide variety of 'climate forcings' as natural factors inducing climate change over time bringing in warm and cold periods, change in temperature in the past half-century recording 'eleven of the twelve warmest years' since 1850 occurred between 1995 and 2006, and the Earth having been 'warmed by 0.7°C over the past 100 years, has resulted from increased atmospheric concentration of CO_2 and other greenhouse gases. It is, nevertheless, very interesting to observe that: "This natural 'greenhouse effect' is what keeps our planet habitable; without it the planet would be 30° colder".[23]

The current rapid rate of CO_2 concentrations increasing since pre-industrial period without having any significant effect on rising temperature for a long time (5,00,000 years) but inducing climate change now is related to 'two great transformations' in energy-use, the first of which began with water power replaced by coal for power generation initiating the Industrial Revolution, while the 'second great transformation' happened 150 years later when 'harnessing oil to the internal combustion engine in the early 20th century' ushering in another revolution. The simultaneous burning of coal, oil and gas hugely adds to the release of greenhouse gases which with accompanying rise in temperature enhance the pace of climate change.***

Table 2.4: Greenhouse Gas Emission from Reservoirs in Brazil and Canada Compared with Fossil Fuel Emissions

Project	**Reservoir area at operating level (km^2)**	**Average Annual Generation per km^2 (Gwt/yr)**	**Annual CO_2 Emission per km^2 (t/km^2/yr)**	**Annual Methane Emissions per km^2 (A/km^2/yr)**
Balbina Dam (Brazil)	3,147	970	7,550	45
Tucuruf Dam (Brazil)	2,247	18,030	4,210	40
Grande Rapids Dam (Canada)	1,200	1,700	190-200	48
Churchil/ Nelson Hydroscheme (Canada)	1,400	16,000	190-200	4-8

Source: Patrick McCully, 1998, *Silenced Rivers: The Ecology and Politics of Large Dams*, Orient Longman. Table 5.3: p. 159 (Taken from P.M. Fearnside, *Hydroelectric Dams in Brazilian Amazon as Sources of Greenhouse Gases, Environmental Conservation*, Vol. 22, No. 1, 1995; J.W.M. Rudd *et al.*, *Are Hydroelectric Reservoirs Significant Sources of Greenhouse Gases*? Ambico, Vol. 22, No. 4, June 1993.

> By 2030 greenhouse gas emissions are set to increase by between 50 and hundred per cent above 2000 levels. Meanwhile, the capacity of the Earth's ecological systems to absorb these emissions could shrink. This is because feedbacks between the climate and the carbon cycle may be weakening the absorptive capacity of the world's oceans and forests. For example, warmer oceans absorb less CO_2 and rainforests could shrink with higher temperature and reduced rainfall.[24]

Incidentally, Philip Fearnside of the Brazilian National Institute for Research in Amazonia provides another source of global warming. According to him, water reservoirs behind dams take a great role in greenhouse gas emissions of terrifying proportions. He cites cases of four such dams in Brazil and Canada in support of his claim, presenting estimates calculated to show the actual role of the dams in hastening up climate change through the addition of greenhouse gases on a dangerous scale (Table 2.4).

> Fearnside calculates that in 1990 (six years after Tucuruf started to fill and three years after the gates were closed at Balbina) Tucuruf reservoir emitted 94,50,000 tonnes of carbon monoxide and 90,000 tonnes of methane; Balbina emitted 2,37,50,000 tonnes of carbon dioxide and 1,40,000 tonnes of methane. Combining the effects of the two gases, Fearnside estimated that Tucuruf had 60 per cent as much impact on global warming as a coal-fired plant generating the same amount of electricity – but 50 per cent more impact than a gas-fired power station. Balbina reservoir, meanwhile had *26 times more* impact on global warming than the emissions from an equivalent coal-fired power station.[25]**

Among the mitigating approaches for checking climate change with an understanding that the ceiling of temperature rise should essentially be restricted to 2°, the first major step is to drastically curtail fossil-fuel related CO_2 emissions. In this regard, 'increased energy efficiency, reduced demand for carbon-intensive products, changes in the energy-mix'[26] are suggested to be very important mitigating steps. In particular, since coal is cheaper and more widely available than other two fossil-fuels – oil and natural gas – its use is more popular than the other two fossil-fuels and with support from very strong political lobbies coal figures much more prominently as a fuel in major emitter countries like China, Germany, India and the United States.

We need to be extremely cautious about the present scenario so that in order to attain our target of reducing emissions we have to keep very strictly to the schedule of 50 per cent in greenhouse gas emissions worldwide 'by 2050 against 1990 levels though in practice energy-related emissions are likely to increase between 2004 and 2030 by 15 Gt CO_2, or 55 per cent'.[27]

Perhaps, one of the most encouraging as well as quite discouraging predictions related to the impact of climate change on agricultural production is that it is likely to have a boosting effect on agriculture in developed countries increasing their productivity, while production in developing countries will decrease, making them dependant on food-imports from developed countries. It is predicted that agricultural potential will increase by 8 per cent due mainly to longer growing seasons while in the developing countries it is likely to decrease by 9 per cent, hitting hard particularly the Sub-Saharan Africa and Latin America.[28]

Undoubtedly, however, the most important natural element that helps agricultural production is soil, the scarcely six inches thick layer of this formed over a long stretch of geological space which has the most important role with its age-long cohabitation with different nutrients as well as insects getting used to indigenous seeds more comfortably rather than imported, genetic seeds bereft of the in-built quality of fighting local conditions. But what is, perhaps, far more important is that the soil *in situ* is not blown away when left without vegetation or crops in particular during gestation periods. Besides, overploughing also loosens up the soil and left without any vegetation cover for a long time gives rise to dust storms which, in fact, have turned out in many regions to become a chronic feature especially for rangelands supporting '3.4 billion heads of cattle, sheep and goats' (L.S. Brown, 2012), losing vegetation by becoming highly vulnerable to wind erosion. In India, too, dust storms in many areas, locally called 'Andhi' (without visibility) is going to become a common feature in the dry heartland of the country. It is a process of desertification at a slow pace and countries like China are going to face the most severe rate in this respect having a large herd of 285 million heads of sheep and goats compared to 9 million in the USA, the former having land fast stripped of any vegetation, and giving rise to very serious proportions of dust storms which blowing far chokes the people of South Korea who almost suffocate for breath over a continuously longer period, from an average of 39 days to 118 days between 2000 and 2011 (L.S. Brown, 2012). Perhaps, none compares with Africa in this respect.

> Among the countries most affected by topsoil loss *via* dust storms are Niger, Chad, northern Nigeria and Burkina Faso.

> In Mauritania, in Africa's far west, the number of dust storms jumped from 2 a year in the early 1960s to 80 in 2004.The Bodélé Depression, a vast low-lying region in north-eastern Chad, is the source of an estimated 1.3 billion tons of dust a year, up tenfold from 1947.... Dust storms leaving Africa typically travel west across the Atlantic, depositing dust in the Caribbean. The 2-3 billion tons of fine soil particles that leave Africa each year in dust storms are slowly draining the continent of its fertility and hence its biological productivity.[29]

In response to the Kyoto Protocol, 1992, the European Union committed in 2007 to reduce greenhouse gas emission 'at least' 20 per cent by 2020 and 30 per cent by 2030 with a further promise to set the goal on a reduction of 60-80 per cent provided an international agreement is reached.[30]

Another major success as a mitigating step achieved was use of 'compressed natural gas' (CNG), reducing dependence on imported oil which had been almost obligatory in India and Pakistan improving cost on fuel, thereby improving air quality and thus public health. However, the most important technological development, known as CCS, or, in other words, Carbon Capture and Storage forms another step in this direction.

> Using CCS technology, it is possible to separate the gas emitted when fossil fuels are burned, process it into liquified solid form, and transport it by ship or pipeline to a location – below the seabed, into disused coalmines, depleted oil wells, or other locations – where it can be stored. Applied to coal plants, CCS technology offers the potential for near-zero CO_2 emssion.[31]

A demonstration of the feasibility of these operations has already been proved in both through partnership between the European Union and the United States but the major snag in making the practice profitably operational is that it requires agreement at the international level on its storage under the seabed. CCS plants planned to come into operation by 2015 were capable of collectively saving only about '15 mt CO_2' in emissions or 0.2 per cent of coal-fired power emission which is hardly any challenge to climate change.

Thus, as it appears, CNG technology is not going to benefit mankind in a big way in cutting down emissions effectively as a check to further rise in temperature and thus slowing down the pace of climate change. Hence, it looks as if there is no alternative to bringing down the consumption of fossil fuels which have proved to be an extremely harmful practice apart from turning to other sources of power generation, such as from wind, sun, sea-waves, tectonic sources, where available.****

It is further pointed out in the UN Report 2007/2008 that in the residential sector application of 'a wide range of policies in building standards, procurement regulations, appliance standards and energy-efficiency obligation.'[32] This makes it possible for '29 per cent saving in emissions' by 2020 while electrical appliances, such as bulbs etc., are capable of using energy more efficiently 'producing a lower carbon footprint', and used in OECD countries from 2005 onwards it would save 'some 322 million tonnes of CO_2 emissions by 2010 and by 2030, this will decrease emission of CO_2 by 572 mt a year equivalent to removing 200 million cars from the road or close 400 gas-fired power situations' (UNDP 2007/2008).

In the world where 1.6 billion people have no access to electricity (UNDP 2009/2008), its production from using rapeseed oil, soyabean, sugarcane and even wheat at some instances to produce ethanol as a substitute petroleum for running cars is no solution, but rather a crime to keep over a billion people hungry as well as malnourished in the world apart from adding more CO_2 emission. For example: 'The 136 million passenger cars on United States' roads contribute 35 per cent of national transport-based greenhouse gas emissions and the 87 million light trucks another 27 per cent'.[33]

It is rather unfortunate to watch the growing support in many rich countries for the production of ethanol which instead of indirectly cutting CO_2 emissions by way of replacing, or, rather cutting down the consumption of petroleum uses grain corns and other types of food, is increasing their consumption from 1 million ton in 1980 to 41 million ton in 2005, scaling up to 127 million ton, one-third of which was in the USA alone in 2011 capable of feeding 400 million people (L.S. Brown, 2012). According to Lester Brown, the grain required to fill a 25-gallon fuel tank of a sporting car could feed one person round the year. Following the United States, Brazil is second in the race for the production of ethanol, jointly contributing 87 per cent of world output followed by China, Canada, France, and Germany, all together producing 23 billion gallons in 2011, having increased from hardly 1 million ton in 1980. The resulting rise in food prices as more food is diverted to production of biofuel is creating a crisis in food-availability among the poor countries which will materially lead to food riots as it happened in some 60 counties between 2007 and 2009.[34]

Plainly speaking, there could be no better substitute for travelling within the urban areas based on public transport and could be made much more effectively done by walking manageable distances or by cycling. Such an attempt would have a very strong effect on cutting on CO_2 emissions. In fact, the bicycle has been described as an 'ideal antidote to pollution' (L.S. Brown, 2004/Reprint). 'It alleviates congestion, lowers pollution, reduces obesity, increases physical fitness, does not emit climate-disrupting carbon dioxide, and it is affordable to billions of people who cannot buy an automobile'.[35]

Besides, it has a parking advantage, 20 bicycles can ideally fit into a road-space used by one car, thus having a tremendous impact on urban land-use. The growing popularity of the bicycle as a means of urban transportation is evidenced in China where its annual production increased from 6 million a year in 1976 to 40 million a year in 1998 turning a total of almost 540 million people into bicycle owners. This growing importance of the bicycle has also been acknowledged by many countries like the USA where over '80 per cent of police departments' use bicycles for routine-checks. In New York, for instance, bicycle messengers are competing for '$700 million dollars' worth of business each year'. In order to recognise the overwhelming importance of the bicycle in the growing transport system, the relevant urban centres should make extra-provisions for providing bicycle lanes uninterrupted by major roads, coordinate the traffic system giving uninterrupted passage for them through the city. Leaders among other developed countries in creating a bicycle friendly transport system are the 'Dutch, the Danes and the Germans'. In this respect, as reported by L.S Brown (2004, Reprint), bicycles enjoy the advantage over motorists in getting right of way at traffic-lights to move ahead of the motorists to get the clearance. The cyclists comprise '30 per cent of all urban trips' in the Netherlands. Among other advanced countries Japan and Spain are making similar efforts.[36]

All the facts and observations covering a wide field are presented to make an idea about the success or even failures in making the food-productive system likely to face the great challenge not only in making the attempt at providing food for the entire global population with wide variations in their choice of food. The rising purchasing power of a growingly larger section of the global population increasing in number at the same time is going to be a great challenge, particularly in the face of climate change with no sign of immediate abating.

On the contrary, poor peasant-farmers and not the corporate giants are the only hope in such a controversial situation, who are likely to stick to their professions of producing food not only for meeting their own requirements but also to fulfil the needs of the remaining population, as they have always done out of their professional proficiency adopted as a life-style who love to touch soil, play with it, and coax out food as a sheer habit as long as their efforts are not seriously interfered with. Hence, it ultimately comes down to no resistance from any quarter since blocking peasant-farmers' path to freedom of working in the field and produce food we genuinely need is the only option left for mankind for its survival in the coming years.

NOTES

***World Commission on Environment and Development** (The Brundtland Report). *Our Common Future*, Oxford University Press, Oxford;1987. p.40. 'Sustainable development seeks to meet the needs and aspirations of the present without compromising the ability to meet those of the future. Far from requiring the cessation of economic growth, it recognises that the problems of poverty and underdevelopment cannot be solved unless we have a new era of growth in which developing countries play a large role and reap large benefits ... the concept of sustainable development will necessarily work to assure that growing economies remain firmly attached to their ecological roots and these roots are protected and nurtured so that they may support growth over the long term. Environmental protection is thus inherent in the concept of sustainable development, as is a focus on the sources of environmental problems rather than the symptoms. *Source*: as above.

****Meadows, D.H., Meadows, D.L., Panders, J., Behren III, William, W.** *The Limits to Growth. A signet Book from American Library*. Times Mirror; 1972. p. 58.
'Recent costs (around 1972) of developing new land have ranged from $215 to $5,275 per hectare. Average cost for opening land in unsettled areas has been $1,150 per hectare. According to FAO report, opening more land to cultivation is not economically feasible, even given the pressing need for food in the world today.'

*****Magnuson, Joel.** *The Approaching Great Transformation: Toward Liveable Post Carbon Economy*. Policy Press at the University of Bristol; 2013. pp. 106-107.
'Americans sustain a daily diet of about 21 million barrels of oil, 2.7 million tons of coal, and 63 billion cubic feet of natural gas. Currently, over 84 per cent of our total energy use comes from fossil fuels, another 8.5 per cent is from nuclear, and only 7.3 per cent comes from renewable sources ... Consider trying to replace gasoline with ethanol. Right now, Americans consume about 140 billion gallons of gasoline every year. Gallon for gallon, the energy density of ethanol is less than gasoline, so getting the same energy and work would take more, about 210 billion gallons of ethanol.... On average ethanol facilities can produce about 41.7 million gallons a year. To get 210 billion gallons of

ethanol would require 5,000 new processing plants and would require growing 74 billion bushels of corn. This would be nearly six times the record corn output of 13.2 billion bushels. It would also require 368 million acres of the best farmland which constitutes about 90 per cent of the total cropland in the United States of which less than 13 per cent is irrigated. Making ethanol out of corn requires an enormous amount of water. As it takes about 1,000 gallons of water to produce one gallon of ethanol, converting from gasoline to ethanol would take 210 trillion gallons of water. That would be about 4.5 times the total amount of surface and ground water currently used in agriculture.'

****Human Development Report 2007/2008.** *Fighting Climate Change: Human Solidarity in a Divided World.* p.135.

'Both wind and Solar Power are expanding sources of global energy. Global investment in renewable energy has increased rapidly, from US$ 27 billion in 2004 to US$ 71 billion in 2006 alone. Remarkable efficiency gains have been registered. Modern wind turbines produce 180 times more energy at less than half the cost per unit than turbines 20 years ago. Investments in the United States have increased wind capacity by a factor of six in the intervening period. Much the same has happened in solar power. The efficiency which photovoltaic cells convert sunlight into electricity has climbed from 6 per cent in the early 1990s to 15 per cent now, while their cost has fallen by 80 per cent … Germany's Renewable Sources Act is an example. This has been used to fix the price of renewable power for 20 years on a sliding scale … In Spain, the Government has used a national permission tariff to increase the contribution of wind power. This now meets around 8 per cent of the country's electricity demand, rising to more than 20 per cent in the densely populated provinces of Castilla-La Mancha and Galicia. In 2005 alone, the increase in wind turbine capacity in Spain saved around 19 million tons of CO_2 emissions'.

Brown, Lester R. *Plan B 2.0: Rescuing a Planet Under Stress and a Civilisation in Trouble*. Earth Policy Institute; 2006. pp.197-203.

Incidentally, we can derive some more information in this regard from another *source:* Lester R. Brown, 2006, 'Plan B 2.0: Rescuing a Planet Under Stress and a Civilisation in Trouble', in which he points out the importance of a few more renewable sources of power playing a great role in slowing down the process of climate change. One of the most important among those sources of power is geothermal energy connected with radioactivity within the earth available either through conduction or through hot springs and geysers as an inexhaustible source almost free of emission. Countries bordering the Pacific in the 'so-called Ring of Fire' are rich in geothermal energy, include Chile, Peru, Ecuador, Colombia, all the Central American countries, Mexico, the Western United States and Canada on the eastern ring whereas those on the western ring consist of Russia, China, South Korea, Japan, the Philippines, Indonesia, Australia, and New Zealand besides those along the Great Rift of Africa and the Eastern Mediterranean.

The most popular use of geothermal energy is for generating electricity and 'to directly heat buildings, greenhouses, aquacultural ponds besides providing heat for industrial activities'. First discovered in Italy in 1904, its use spread fast to a global capacity of 8,400 megawatts by 2003. According to Lester R. Brown (2006), the USA generating 2,000 megawatts followed by the Philippines with 1,900 megawatts, lead the world, providing 27 per cent of electric supply in the latter. Among other leading producers are Italy, Mexico, Indonesia, Japan, and New Zealand. More than thirty countries tap this power for heating purposes and among them Ireland and France are heated by geothermal power, and thus saving $100 million spent on oil-imports and, as it looks, Ireland is going to be the first country in the world to get out of the use of fossil-fuels. Another highly important use of geothermal power in cold countries like Russia, Hungary, Ireland and parts of the United States is heating greenhouses for the production of fresh vegetables in winter while, on the other hand, its use for aquaculture in countries like China, Israel, and in the United States provides some varieties of fish with a quick growth (Tilapia, Striped Bass, Catfish). Above all, such multifarious use of this power obtained free from geothermal energy had a very wide use for bathing and swimming by heating up the water already developed on a grand scale both in Japan and Ireland, while Indonesia plans to get all its energy-supply from geothermal sources.

In summarising the potential of leading industrial as well as some of non-industrial countries in replacing the fossil-fuels with renewable energy the United States possesses immense possibilities with rich endowment of low-cost wind energy likely to emerge as the 'centre-piece' of the energy economy' which can meet all sorts of industrial requirements including 'steel melting'. This, along with immense hydel potential and the availability of 'substantial geothermal potential and enormous solar cell potential', puts the country in a very strong position for contributing effectively to cut carbon emission.

Similarly, Germany is going ahead with a determined approach to cut carbon emission by 37 per cent by 2050, through raising energy efficiency while another 45 per cent from renewables using wind and solar power. Indonesia, as mentioned above, possesses immense resources of wind, solar as well as thermal power to get all electricity she needs. In Spain, solar cells and panels as well as wind power are going to take prominent place in power generation. On the other hand,

Brazil is almost at the point of getting free of oil-imports through self-sufficiency attained in sugarcane-based ethanol production. The country has further benefited by a 'generous supply' of hydroelectricity and greater scope in the generation of wind and waterpower. In China hydropower already meets 15 per cent of her power requirement and the country has a great potential for the generation of power from wind which can easily be doubled.

In the United Kingdom, wind power generated primarily from off-shore winds has immense possibilities while combined with wave-power along with solar heating the country enjoys an enviable position. In Argentina, on the other hand, hydropower takes a very important place in meeting her power requirement supplying alone as much as 42 per cent while being one of the richest wind resources, particularly in Patagonia, she also has great potential in Solar Power. As Lester R. Brown (2006) observes, it is highly interesting to note that in the 20th century, the Middle East with its vast oil-resources had a very strong grip on the global economy giving rise to 'globalisation of the energy economy' in sharp contrast to an emerging reversal of its 'localisation' depending on the use of local power resources.

1. Meadows, D.H., Meadows, D.L., Panders, J., Behren, III, W., W. *The Limits to Growth: A Report for The Club of Rome's Project on the Predicament of Mankind.* New York: Universe Books, 1972, pp. x, xi.
2. *Ibid.,* pp. 57-58
3. Meadows, D.H., Meadows, D.L., Panders, J., Behren, III, W.,W. *The Limits to Growth: A Report for The Club of Rome's Project on the Predicament of Mankind.* New York: Universe Books, 1972, pp. 58-59.
4. *Ibid.,* pp. 62-63.
5. *Ibid.,* p. 199.
6. Meadows, Donella H. "Just So Much and No More: Let Us Rejoice in the Logic of Earth's Economy", in *Resurgence,* 210. 2002. p. 33.
7. Helm, Dieter. *Natural Capital: Valuing the Planet.* New Haven and London: Yale University Press, 2015, pp. 24-25.
8. *Ibid.,* pp. 26-28.
9. *Ibid.,* p. 36.
10. *Ibid.,* p. 37.
11. *Human Development Report 2007/2008***.** Fighting Climate Change: Human Solidarity in a Divided World. United Nations Development Programme, pp. 3, 7.
12. *Ibid.,* p. 9.
13. *Ibid.,* p. 25.
14. *Ibid.,* pp. 27, 30.
15. *The United Nations World Water Development Report.* Water for People Water for Life. 2003, pp. 10-13.

16. Brown, Lester R. *Full Planet, Empty Plates*. New York, London: W.W. Norton & Company, 2012, pp. 57-59.
17. *Ibid.*, p. 70.
18. *The United Nations World Water Development Report*. Water for People Water for Life. 2003, pp. 76-78.
19. McCully, Patrick. *Silenced Rivers: The Ecology and Politics of Large Dams*. Hyderabad: Orient Longman, 1998, p. 119.
20. *Ibid.*, p. 124.
21. *The United Nations World Water Development Report*. Water for People Water for Life. 2003, p. 84.
22. *Human Development Report 2007/2008*. Fighting Climate Change: Human Solidarity in a Divided World. United Nations Development Programme. p. 31.
23. *Human Development Report 2007/2008*. Fighting Climate Change: Human Solidarity in a Divided World. United Nations Development Programme. p. 33.
24. McCully, Patrick. *Silenced Rivers: The Ecology and Politics of Large Dams*. Hyderabad: Orient Longman, 1998, p. 160.
25. *Human Development Report 2007/2008*. Fighting Climate Change: Human Solidarity in a Divided World. p. 51.
26. *Ibid.*, p. 55.
27. *Ibid.*, pp. 90-91.
28. Brown, Lester R. *Full Planet, Empty Plates*. New York and London: W.W. Norton & Company, 2012, pp. 50-51.
29. *Human Development Report 2007/2008*. Fighting Climate Change: Human Solidarity in a Divided World. p. 113.
30. *Ibid.*, p. 145.
31. *Human Development Report 2007/2008*. Fighting Climate Change: Human Solidarity in a Divided World. p. 138.
32. *Ibid.*, p. 139.
33. Brown, Lester R. *Full Planet, Empty Plates*. New York and London: W.W. Norton & Company, 2012, pp. 36-40.
34. Brown, Lester R. *Eco-Economy: Building an Economy for the Earth*. Hyderabad: Orient Longman, 2004, p. 199.
35. *Ibid.*, pp. 199-201.

3
Are the Earth's Resources Sufficient to Sustain Global Farming Practices?

I. Playground of Food Corporate Houses, Food Waste and Rising Prices

Looking back at the availability of food, which became scarce since 2007/2008, the UK provides one of the glaring instances of food scarcity where food price rose 12 per cent in the five years since 2007 and the 'poorest tenth' of the British people had to spend 17 per cent more on food in '2011 compared to 2007', depending very much on charitable assistance with more than 320 food banks working in the UK 'issuing food parcels to over quarter of a million people' in 2012. In the UK, the proliferation of supermarkets providing processed or semi-processed food ousted the small retailers giving a wider choice in the selection of food. The processed food naturally was much less tasteful than the variety of fresh food offered by corner-shops but they started to slowly close down in the stiff competition from the mushrooming Tesco (international), Sainsbury (British), Morrison (British), Asda (American), Aldi (German), Walmart (American), Waitrose (British), Lidl (German) swarming the country offering food at competitive prices unlike the corner shops with a variation of fresh food-prices.

The processed food, only needing to be reheated at home, unavoidably gave rise to wastes and this has been the nature of development all over Europe of a growing supermarket culture taken up by large corporate houses like Tesco ousting the corner-shop from business, the latter offering a wider choice in fresh food. The ultimate result of this change from cooked food to processed, or even semi-processed food is lack of tasty food leading to some of it left un-tasted on the plate and thrown into the bin. The enormity of this as well as the severe consequences of such waste may be ascertained from an assessment of the European Commission that altogether 50 per cent of edible food is wasted with far-reaching consequences. According to them: 'Food waste is equivalent to 15 million tonnes of CO_2 emission, and the vast majority goes into landfill, where it generates methane'.[1]

In the background of this colossal wastage of good food, it is very much shocking to take note of the observation of Fitzpatrick (2014) who thinks that if globally about one-third of edible food equivalent to 1.3 billion tonnes is wasted every year due to several reasons including damage during harvest, transport, storage or their cosmetic value (appearance or how they look), an amount which could feed the 'world's hunger several times over'. The situation for the hungry or underfed gets much more serious if, as observed by Oxfam, the export prices rise expected between 2010-30 by '177 per cent for maize, 120 per cent for wheat and 107 per cent for processed rice' partly due to climate change, and unavailability for global population (likely to reach 8 billion by 2030), and food is exported rather than being available locally, and reliance on animals goes on rising. 'It takes an estimated 11 times as much grain to feed a man if it passes through a cow first ... It also takes a staggering *thousand* times more water to produce a kilo of beef than wheat'.[2]

II. The Status of Small Farmers

After taking a bird's eye-view of one of the leading industrial countries, the UK, in terms of availability of food, food-waste along with reasons behind the worsening situation over the years, we may turn our attention to the real food-suppliers living in the rural areas of the developing world. More than half of the 3 billion rural people living in areas with commercial agricultural potential close to urban areas have 'exceeded expectations in food production'. The World Bank Report, 2003, made some very important recommendations for strengthening the system of rural agricultural producers who, as predicted in the Report, were expected to bear the responsibility of feeding an additional 2-3 billion people in the coming 30-50 years (from 2003) which necessitated some desired but essential steps to be taken ahead adding to the capability of farmers in taking up such a great responsibility. The most important change in the institutional rules comprising moving the land assets into the hands of the rural farmers, which in other words, mean give the farmers rights to land they operated on along with taking such other essential steps to make agricultural process of the farmers operative. They consist of 'halting nutrient mining, reducing soil erosion, and adopting agricultural practices that restore soil fertility. It will also require sharing land and water to serve the expanding urban population and meet environmental needs'.[3]

According to this Report (The World Bank, 2003), agricultural expansion till then has taken a heavy toll on global biodiversity through clearance of one-fifth of tropical forests since 1960, making the existence of remaining biodiversity precarious, over one-third of it confined to 1.4 per cent of the global land-surface. Among other restraining factors are the land getting increasingly degraded and water being scarce. On the contrary, however, according to agricultural economists,'the food problem is the result of 'insufficient' purchasing power of the people'. A change in consumption pattern – now dominated by meat and poultry in OECD countries along with China – contrasted with eliminating hunger among the poor people living in the countryside depends on 'increasing their ability to produce food to consume and food to sell in the markets'.

There are, however, factors restricting an increase in food production and they are erosion, salinisation, compaction as different forms of soil degradation affecting 30 per cent of the world's irrigated lands, 40 per cent of rain-fed agricultural lands and 70 per cent of the rangelands, resulting in loss in overall global production by 12 per cent altogether over three decades, coming to an average loss of productivity by 0.4 per cent. Interestingly, 'time series data for 1971-73' from India and Pakistan indicate that intensification of land and water use has caused resource degradation decreasing production.

Among the factors likely to interfere with farmers' production, water poses serious problems as according to the prediction of World Commission on Water, its use will increase putting 4 billion people under severe stress by 2025, naturally affecting agriculture harder in competition with urban areas. In this respect, China is already exploiting 44 per cent of its useable water projected to be increased to 60 per cent by 2020, which according to experts exceeds 'environmental carrying capacity'. Similarly, water use in Afghanistan already 'exceeds environmental limits' while this is also going to be the case in Pakistan and India by 2020. In contrast, Africa with hardly 1-3 per cent of her agriculture under irrigation, compared with two-thirds in Asia, has limited potential as 60 per cent of this lies in the humid regions with high rainfall.

Similarly, land, the other constituent of food production also appears to slowly be decreasing in availability getting lost globally through degradation at the rate of about '0.5 per cent per year and another 0.1 per cent annually through new infrastructure'. Water erosion severely degrades land in South East Asia and Latin America, soil nutrient mining degrades land in Africa while salinization is a great problem for land degradation in irrigated areas. Almost '303 million hectares of land has thus been lost in Africa and Latin America where 307 million hectares are now available for cultivation'. Against this background, however, both Latin America and Africa together contain almost 1,400 million hectares of tillable soil but all of this is in forest which becomes a very expensive exercise to clear for agricultural use.

However, if one considers the most important problem of rural growth with its very powerful effect on pulling rural people out of eternal poverty and faced with environmental problem on the one hand as mentioned, and on the other great responsibility of feeding the world and thus rescue it from the grip of corporate houses keen on producing industrial crops. Thus, in order to 'break the poverty cycle' and to take the rural poor out of poverty, two steps need to be taken (i) creation of non-farm jobs in the rural sector; and (ii) provision for training in non-farm employment.

III. Role of Agriculture as an Instrument for Overall Development

As mentioned in The World Bank Report 2003, agriculture plays a very important role in the growth of a national economy, generating an average 29 per cent of the gross domestic product (GDP), employing '65 per cent of the labour force' and agricultural production is the main source of income guaranteeing food security for the majority of the rural population. Second, it is the major source of livelihood for almost '86 per cent of the rural population' providing jobs for '1.3 billion small-holders and landless workers'. Contributing to a fall or reduction in rural poverty from '37 per cent to 29 per cent' between 1997 and 2002, resulting from better conditions in rural areas rather than outmigration. But these changes are largely confined to East Africa and the Pacific rather than in South Asia and Sub-Saharan Africa. On the other hand, its negative role as a despoiler of the environment, as the largest user of water giving rise to water scarcity along with depletion of underground water, its role in agrochemical pollution, soil exhaustion and global climate change 'accounting for up to 30 per cent greenhouse gas emission' almost go unchallenged considering the services it offers in keeping away 'carbon, managing watersheds and preserving biodiversity'. Nevertheless, 'the role of agriculture as a positive factor in development can still be made constructive with careful management and conservation of natural resources in harmony with environmental use as an integral part.'[4]

It is claimed in The World Bank Report 2008 that 'agriculture has special power in reducing poverty', and not the industrial agriculture of corporate making. 'For China, aggregate growth originating in agriculture is estimated to have been 3.5 times more effective in reducing poverty than growth outside agriculture – and for Latin America 2.7 per cent more'.[5]

But the author very strongly disagrees with the contention that India gained from the 'Green Revolution Movement' – as popularly known – using imported seeds instead of indigenous ones acclimatised with local conditions through use for generations. The severe cost the country had to bear soon after with complete loss of its biodiversity as well as local seeds because the imported seeds gave dwindling returns in the production of crops for wealthy farmers while leaving poor farmers deeper in debt.

It does not help poverty reduction, if the farming practices can be a leading factor as a precursor of industrial growth, as it happened in Britain and many other countries including the USA where as well as stabilisation of agriculture made way for the industrial use of crops produced. More recently China, India and Vietnam provide such instances of this economic reversal.

However, before taking up the brighter side of farming practices and advocating support to its more secure passage we may turn our attention to the damaging role it plays enhancing the pace of climate change, or the overall disaster mankind awaits in the future.

According to Wayne Roberts (2013, New Edition) 'Food production is the world's most polluting industry, doing more damage to more territory than logging, mining or heavy industry. The environmental damage it does include over-fishing, soil degradation, excessive irrigation, and reduction of genetic diversity among farm animals and plants all of which are direct threats to future food security, not just to other creatures that share the planet with us'.[6]

Such an extremely discouraging statement is based on the facts that, as stated by him, to meet the growing meat-intensive diet of the Global North agriculture needs, '700 million tonnes of fertilizer-intensive grains and 200 million tonnes of pesticide-intensive soyabeans to feed world's 27 billion livestock' responsible for 'overloading waterways with manure and clogging the atmosphere with global warming methane gas from animal farts and burps' (Wayne Roberts, 2013).

Another darker side of the world agriculture forming the world's largest labour force consisting of about 1.1 billion representing farmers, farm workers and fisherfolk besides employing 170 million child labour, more than any other industry. But what is more interesting is that:

> an authoritative 2013 report by Oxfam says that 450 million farmers who supply the 10 biggest food companies account for 60 per cent of the world's poor and 80 per cent of its hungry ...

> A multi-trillion dollar-a-year food industry selling a necessity of life impoverishes more people than any economic sector.[7]

This highly contrasting situation swelling the number of hungry people over a billion, a 25 per cent increase since the mid-1990s, put them in a combating mood, leading 'peasants, small-scale farmers, farm workers and indigenous communities' to organise a transitional agrarian movement, called La Via Campesina, formed in 1993 representing 148 organisations from 69 countries raised voices of:

> radical opposition to the globalisation of an industrial and neoliberal model of agriculture, claiming that time for food sovereignty aims to transform dominant forces, including those related to politics, economics, gender, the environment and social organisation, there will, no doubt, be a long and hard struggle to see food sovereignty become the standard model for food production and rural development.[8]

In other words, as pointed out above in the preceding paragraph, food sovereignty is claimed as 'more than a right'. It involves a fight against 'literal displacement of millions of farmers from the land, the expulsion going on with unabated pressure of urban expansion at the cost of living space as well as the only way of living for them. It is not only the struggle for livelihood but also for obtaining the fundamental right to the entitlement of living as a human being. According to Altieri (2010), 'global forces are challenging the ability of developing countries to feed themselves'. The developing world is facing this challenge in the form of development of industrial agriculture at the initiative of very powerful corporate houses often based on a monoculture, such as soyabeans in Brazil, which deriving economic benefit from its sale, remain satisfied ignoring how food crops for direct human consumption is consumed elsewhere (in buyer's country) for enhancing meat production, say in China, and thus making a severe impact on the economy of local farming communities dislodged from their practices.

In this respect, the society needs to be much more conscious about the ulterior motive of the 'growing push towards industrial agriculture and globalisation' characterised by greater emphasis on expansion and production of such biofuel crops like sugarcane, maize, soyabeans, oil palm, eucalyptus etc. and thus 'reshaping the world's agriculture and food supply' with serious effect on the character of the crop productivity in the tropical zones of the developing world.[9]

In sharp opposition to this self-destructive system the food-growers, particularly representing the small farmers engaged in farm practices traditionally for generations were quick to realise the importance of applying 'ecological concepts and principles to the management of sustainable productive systems. This has been proved to be an immensely successful approach to a safe and secure food production system providing healthy food not only to the family of growers themselves and their families but also take up the great challenge of feeding the growing population in the coming years, even maybe beyond 2050 when the world becomes more urban than rural with over 66 per cent of the global population living in urban areas (UN: 2014 Revision of the World Urbanisation Process)'.[10*]

It is very strongly believed that the built-in resilience of small farm holders capable of adjusting to 'rapidly changing climates, pest and disease' is the most important factor in agricultural practices keep going tidying over any resistance. This is the real secret behind the success of traditional farming still now capable of providing food for this world getting hungrier in the lack of it under the growing pressure of industrial agriculture. This amazing feat on the part of small rural farmers highly appreciated by Altieri (2010) is expressed in the following words:

> The persistence of millions of agricultural hectares under ancient, traditional management in the form of raised fields, terraces, polycultures (with a number of crops growing in the same field), agroforestry systems etc., documented a successful indigenous agricultural strategy, and constitutes a tribute to the *creativity* of traditional farmers.[11]

As Altieri points out, luckily there are thousands of small traditional farms still functioning in the Third World which maintaining agroecological approaches take an active role in producing food and in order to keep their activities going on in full strength gave rise to the concept of food sovereignty putting emphasis on farmers' right to land, access to seeds, the indigenous one, as well as facilities of local autonomy, local markets, local production-consumption cycles, energy, technology as well as farmer to farmer contacts.

The basic idea behind food sovereignty is food-security, which in Wayne Roberts' (2010) opinion is that 'all people at all times have access to adequate amounts of safe, nutritious and culturally appropriate food'.

Founded in 1992, La Via Campesina, roughly meaning the peasants' or country peoples' way, representing some 200 million peasants, agricultural workers and indigenous people from some 70 countries, food sovereignty took form as a way of 'mass improvement of peasants and fisherfolk in the Global South' demanding the restoration of trade rules prevailing before 1990s exempting food and agriculture from free trade deals by WTO (World Trade Organisation). Wayne Roberts (2010) tries to clarify the *actual relationship* between the two words – food and sovereignty – which, for the first time, taken together has a different connotation for the Global South containing fundamentally rural countries populated by peasants, farmworkers and fisherfolk. They, thus, differ from the food-growers of the Global North treating their work in agriculture as a business while in the Global South they treat their work by non-monetary values, making their interests as the same as of the people of the Global South.[12] It is highly relevant in this connection to emphasise the major principle leading to this point of difference between the Global North, which considered agriculture as business and the Global South taking food production as means of feeding themselves as well as feeding others as a bounden responsibility. To be specific:

> Food sovereignty comes from the poorest of the poor peasants and farmers who, among other things, faced bankruptcy as a result of competition with subsidised food staples from Europe and North America. If food security spells out the rights of consumer sovereignty, food sovereignty corrects that imbalance by reasserting the primacy of production ... It is a side that grows out of deep roots in history, ecology, community, and spirituality. It presents a food system that comes from a different place, a community-based food system.[13]

The most important facet of the global food crisis staring in the face of the growing global population can no longer be ignored or put aside thinking that we are going to get out of the control of the present capitalist agriculture growing fast under the patronage of corporate monopolies undermining the global food system. The experience of the 'record numbers of world's poor going hungry and poor families remaining unfed in 2008 in spite of the record harvest in 2007, as stated by the FAO, when there was more than enough food in the world to feed everyone in 2008 which was enough to feed everyone – at least one and a half times when growth-rate of food was higher than the growth-rate of global population over a period of two decades, 2 per cent against 1.4 per cent for the latter. According to this source, it is lack of buying power of nearly 3 billion people living in the Global South that keeps them hungry, almost half of them earning less than $2 a day while 20 per cent classified as extremely poor earns less than a dollar a day.[14] In spite of the high claims of the Green Revolution, hunger spread like fire bringing over 1 billion people under its ambit by 2010 when the author Eric Holt Giménez was writing this report, pointing out further that when 50 years ago the developing countries had 'yearly trade surpluses of $1 billion, the decades of capitalist expansion of industrial agri-food complex has given rise to food deficit in the Global South rising to $11 billion'.

According to him, 'the capitalist agricultural expansion' during the last half century has constantly dispossessed the world's peasantry of land, water and genetic resources such as indigenous seeds with their endless varieties in the garb of development and colossal benefit derived from such fraud amounted to making three seed companies – Monsanto, Dupont and Syngenta – monopolise 39 per cent of the global seed market.

However, the most interesting thing is that in spite of such bulldozing pressure of these giant corporates of industrial agriculture, the peasant farmers resisted being ousted from their family occupations keeping their absolute numbers 'remarkably stable over the last forty years, providing significant amount of food in the Global South', as much as 90 per cent of the food required in African countries.[15]

This quite naturally leads us to the next question, such as what had been the mechanism for the peasant farmers spread widely over the developing countries, or in the Global South as it is called, to survive the determined, brutal onslaught in different guises even with the support of global organisations like WTO (World Trade Organization). This, in fact, led to the formation of 'peasant-based call for food-sovereignty' with the idea of 'people, rather than corporate monopolies' in making decisions for human food. Food sovereignty, in fact, carries a much deeper meaning than food security, guaranteeing not only access to food but also assuring a 'democratic control over the food system', encompassing all the relevant stages from 'production, distribution, marketing and consumption'. Besides, the entire process has not only acknowledged but also attached great importance to the role of women in resisting any pressure from any quarter in replacing native seeds with imported ones, working always very hard in diversifying their crops, protecting soils to preserve their nutrient wealth, 'conserve their water' and forests, in other words, the biodiversity, besides establishing 'local gardens, markets, businesses and community-based food systems'.[16] In other words, this highly protective as well as … productive and resilient strength of rural peasant women has always acted as, perhaps, the most supportive pillar of success of peasant, small-scale farmers deriving strength for survival through many obstacles.

The international gathering at Mons in Belgium in 1993, supported by a number of NGOs made a final verdict against stiff opposition from the Industrial Corporate Houses accepting the moral right of the international peasant movement, *La Via Compesina* asserting the 'right of small farmers to make a living in the countryside, the right of all people to healthy food, and the right of nations to define their own agricultural policies'. In 2006, FAO's International Conference on Agrarian Reform and Rural Development held in Madrid unequivocally accepted the call for 'food sovereignty' as a movement correcting the wrong premises of solving food crisis through industrial agriculture wrongly supported by many International Organisations (IMF, The World Bank, WTO etc.). It also ended the continuation of land-grab policies involved in selling the land of the poor peasants by the relevant governments in league with buyers from outside and also rejected support for projects like Green Revolution upsetting biodiversity etc. for good. This entire approach to correct the wrong done so far by the promoters of industrial agriculture was made by a large group of over 'four hundred scientists and development experts from more than 80 countries, concluding that: 'there is an urgent need to increase and strengthen further research and adoption of locally appropriate and domestically controlled agroecological methods of production relying on local expertise, local germ plasm, and farmer-managed local seed systems'.[17]

The most vibrant response to agroecological transformation as a sequence of 'Campesino a Campesino' or farmer to farmer approach is witnessed in Latin America's thirty-year long movement with a series of rural projects spread over a large area, from the Guatemalan highlands in the early 1970s to 'hundreds of thousands of small-holders across the Americas' achieving the 'most dramatic success' in Cuba, completely transforming its high external input, large-scale agricultural systems to 'smaller low-input farmer-driven agroecological practices creating 'over 1,00,000 small-holders' in eight years while it had taken nearly twenty years for Mexico and Central America to grow to that size. The major aspect of the spectacular success of this movement depended on a clearly defined methodical process which laid great emphasis on the importance of addition of organic matter to the soil including green manures, crop diversification, different methods of 'pest management, biological weed control, reforestation, and agrodiversity at farm and watershed levels', keeping an eye on 'seed selection as well as soil and water conservation'.[18]

IV. Nature's Storehouse of Food and Feeding Capability of Rural Farmers

It is highly interesting as well as important to know that non-field food, that is not cultivated, takes an important place in daily food involving as many as 1.6 billion people who 'manage forests for their livelihood'. As Wayne Roberts (2013) observes:

> forest work has modest start-up costs.... A multiple-use forest can be managed as a grocery, pharmacy, and hardware store without walls, providing security for the very people most likely to face hunger. Berries, mushrooms, fibre for clothes, wood for building, thatch for roofing, found materials for crafts, fuel, medicine, fodder for livestock; it is the foraging version of multi-tasking. The FAO's Dr. Cherukat Chandrasekharan estimated that such renewable non-timber uses of the world's forests contribute $120 billion a year to the economy, on into the indefinite future.[19]

Apart from this, the role of forests in 'non-extractive economies' comprise their environmental services like 'storing carbon, preventing soil erosion and cleaning water filtering down to the water table'.

The role of forests as a storehouse of food for human beings is further corroborated by the fact that man obtains multifarious types of services from the forest free of cost. Man obtains a lot of food and medicines in various forms some of which such as honey are better in food value than man gets from artificial hives when natural ones are rich in antioxidants. Besides, resin collected from trees is made into chewing gum, resulting in a $100 billion dollar industry. And resin especially collected from pines has great value in its use in India serving as a repellent for mosquitos and other insects at night. Among other extremely valuable food-items available from forests are an annual supply of 25,000 tons of nuts from Brazil's Amazonian forests, another 1,300 tons of pine nuts from Turkey along with sorrel and spring leaves of willow used as salad greens while wild spring greens in Canada appear before other crops. Foods with medicinal properties are also available from forests, like Ginseng, yerba mate which is a kind of mineral-packed tea available from the Amazon along with boars and original pigs as food. The forests in Malawi, on the other hand, provide '37 different species of leafy vegetables, two root vegetables, 21 fruits, 23 mushrooms and 14 varieties of edible caterpillars'. In India, the 'Neem' tree has high medicinal value and especially the leaves are eaten raw or cooked as a delicacy while small twigs are very popular for use in brushing teeth all over India for which 'Neem tree' should be considered as a national emblem as peacocks among birds. (This production-link of food-medicine-fibre-fuel makes a great contribution to human requirements in the Global South). Apart from Neem tree, India possesses another tree of great human value in its food-chain grown over a large part of the country which is coconut used for '45 different purposes, from food to fuel to medicine, to fibre for mats and bags' even for making cosmetics. But this nature's storehouse for human (even animals) food etc. obtained free is going to face a bleak future in a short time since world's forests are shrinking at a rate of two per cent a year since 1950. For instance, forest cover in Africa has shrunk from 699 to 635 million hectares between 1990 and 2005, and from 891 to 832 million hectares in South America where they have been mercilessly slashed to give way to the production of palm tree to produce biofuels for replacing oil.[20]

Incidentally, forests are not the only sources of nature's gift of food obtained free. Among other sources meadows and world grasslands provide equally a large variety of 'food, medicinal sources, fuel, fibre and fodder'. Wild plants and herbs often called weeds, some of which are exceptionally nutritious such as dandelions are 'a powerhouse of vitamins A and C, as well as minerals like 'calcium, potassium, silicon and magnesium'. Purslane is rich in iron; stinging nettles are rich in B vitamins while wild groundnut tubers are high in protein. On the other hand, leaves and stems of the wild plants found in open meadow are 'tastiest' in early spring salads and there are also dried seeds very palatable in baked food, along with nutrient-rich seaweeds while a lady in Chile introduced the author (Wayne Roberts) to medicinal teas and creams she makes from plants washed up on the beach and sea vegetables rich in 'manganese', zinc and iodines used in 'soups, salads or puddings'. According to him, *The Cambridge World History of Food* recognises seaplants as 'rich sources of protein, iodine, phenols and fatty acids' taking an important place in Japanese and Korean cuisines. The author (Wayne Roberts) goes further to check the list of eatable and eaten food provided directly by nature and refers to *The Diversity of Life* by Edward O. Wilson saying that there are '30,000 edible species of plants' but only 7,000 of them are used by man when many of the unused plants are 'superior to those in favour', such as the 'delicious lulo fruit in Colombia and Ecuador or the winged bean of New Guinea', and according to him 'people do not die for lack of income ... but lack of access to the wealth of the commons', in other words, 23,000 plants unused by man.[21]

Now, turning to another source, Vaclav Smil, 2002, we get a mine of information about how common people in Africa survive taking recourse to sources of food provided free by Nature herself. One of the most important examples in this regard is provided by wildlife which includes large, widely variant types of species consisting of ants, maggots, snails and lizards, small birds, and rats to hedgehogs and pythons', providing especially seasonally a large share of local diet. For example, another expert, Price (1997) found that wild food gathered particularly in the rainy season from paddy fields comprised nearly half of the total food consumed by farmers in Thailand's Central north-east. Another study (Assibey, 1974) makes claim that 'at least 160 tonnes of meat' of wildlife was available for sale in a single market in Accra, Ghana over a period of 18 months'. Similarly, at least 20 per cent of all animal protein formed part of food in many areas of Africa and Latin America while, as reported by another researcher (Njiforti, 1996) that in Cameroon the 'respondents consumed annually almost 6 kg of bushmeat', representing '24 per cent of the animal protein intake in the region' ('North African porcupine, guinea fowl as the most preferred species'). The same practice is pursued among the rural people in Latin America and South-east Asia and Siberia where 'several hundred million people supplement their food with meat of wild, hunted animals'[22].

Referring to Wilson (1985), Smil strongly supports the view that the natural provision of leaves, seeds, tubers, fruits and nuts together take an important place in the diets of rural population in many areas and one of the 'outstanding' examples is provided by high protein concentrate from the beans of *Parkia biglobosa*, a leguminous tree growing in Western Africa which after 'boiling, fermentation and drying' provides '20-50 per cent of protein with lysine content similar to that of an egg, 30-40 per cent of lipids (three-fifths unsaturated fatty acids)' and plenty of 'iron and calcium'. At the same time *dawadawa* has greater nutritional value, though less popular in use. But use of the powdered leaves of *baobab* and *vigna unguiculata* in stews is quite popular and various African people also extract butter from *shea nuts* and fermented *selerocaryabirrea* fruits to make beer. In Tanzania, nearly half of all meals consumed in rural areas comprise four-fifths of all leafy green, while excellent sources of vitamins and minerals collected are wild. Fleuret (1979), according to Smil, maintains that only 23 out of 114 species of trees and shrubs growing in Sahelian desert zone of Africa are of great importance to human diet while another 46 types are of limited importance in West Africa, 24 out of 165 studied plants are 'regularly eaten but in the continent as a whole more than 500 different wild plants are 'consumed as food'. At the same time, it is highly interesting to know that a 'counterpart of African *dawadawa* is unripe pods of jant (*Prosopis Cineraria*) consumed in Western semi-desert areas in pickled or dried form rich in protein for poor rural folks.[23]

The accounts of dependence on Nature as a source of uncultivated free gifts to satisfy human hunger, mainly in the rural areas, are also available in plenty from countries like Bangladesh and India where folk economy makes a refreshing part of human history. Referring to these two countries with contrasting climatic situations, the meaning of agriculture seems to take a much broader perspective than hitherto treated. As observed in this respect clearly showing unity in diversity:

> On both the food plains of Bangladesh and the dry lands of the Deccan Plateau in India, village women cannot imagine the preparation of food for their family members or raising livestock without leafy greens, grasses, roots, and tubers, leaves of trees, fruits, small fish, molluscs, and other items collected from the borders between fields, along roadsides, in small patches or fallow lands, in ponds and other waterbodies as well as cultivated fields and homesteads. When women do weeding work or return home from grazing animals, they routinely gather these items. They are collected from any space, whether it is their land or not, so long as it can be done in a way that does not interfere with the cultivated crops. From this point of view, the uncultivated plants and animals constitute a prominent place in the local food system and are consequently a vital part of agricultural practice.[24]

The significance of this practice of collecting 'uncultivated biodiversity and using it to prepare food or feed livestock' brings women into a distinct relationship with cultivated and uncultivated space and the gifts of nature, thus elevating as well as recognising the role of women in achieving food security. Besides, being in close relationship with nature – both cultivated and wild – they were the ones to realise the damaging impact of pesticides and herbicides on their and their children's bodies apart from the 'loss of the leafy greens in agricultural fields and fish in the various waterbodies'. The women, in fact, pioneered the campaign against their use and succeeded in bringing an end to the use of pesticides and herbicides.

Those community-based farmer movements involving more than 1,70,000 farm families spread over 15 different districts in Bangladesh at the time of writing this Report by Mazhar *et al.*, (2007), locally called '*Naya Krishi Andolan*' (New Agricultural Movement) based on the development of ecological agriculture was aimed at the development of food policies, farming practices and community relations in combination with the conservation of uncultivated biodiversity along with its use as a source of food for humans and domestic animals. It is necessary to point out here that the uncultivated plants are plucked rather than uprooted helping regeneration in this respect, there is, however, some similarity in the practice where some cultivated plants like mustard and jute from which edible leaves and shoots are collected as an addition to food besides the major cultivated crop. As a matter of fact, farmers consider plucking of leaves during growing season is good for the cultivated crop. Such foods, known locally as 'derived food' plucked during the full-growth period of the major crop include 'leaves of pumpkins, sweet gourd, white gourd' etc., are plucked almost everyday along with flowers of some cultivated crops. In addition, a large variety of fish available in waterbodies, paddy fields, ponds, canals, and rivers with as many as '69 different varieties' documented in one locality (Hinganagar) are highly nutritious, quite often collected by children on their way home while bathing.[25]

Turning to the dry lands of Deccan Plateau in India one finds a different picture, yet here also a large number of cultivated as well as uncultivated biodiversity are used by the people as sources of food. The information here gathered refers to almost 75 villages in the Medak district of Andhra Pradesh representing about a 1,00,000 people composed mainly of small farmers. The land that supports them produces a 'wide variety of agricultural crops', such as sorghum, different varieties of millets, pulses and oil-seeds as rain-fed crops, with a typical farm growing between 8 to 10 crops. However, the number of uncultivated species generally 'most abundant' during the heavy monsoon and those germinating during the later stages of monsoon were ready for consumption during the winter season. Thus, so far identified there are 79 types of uncultivated leafy greens including 'roots, tubers and fruits' ready for consumption as food including edible parts of main cultivated crops, 'such as amaranthus cultivated for its grain' while its leaves are plucked throughout the growing season doing no harm to the plant.

As pointed out in this Report prepared by Mazhar *et al.,* (2007), use of chemicals like herbicides does the greatest harm as no fodder is available for the animals either from the cultivated crops or from the uncultivated plants in the field. On the contrary 'the mixed cropping strategy' practised without the use of any herbicides or pesticides provide a 'continuous supply of green fodder during the cropping season, popularly known as 'weeds comprising stalks, straw and husk of the cultivated crops'. It has been found that in Deccan dry lands 'two acres of mixed crops' support three heads of cattle getting in return not only milk but nutrition for soil in the form of urine and dung used by farmers before ploughing. There is a form of energy recycling for the soil using a type of plough turning only the surface layer of the soil which makes it possible for tiny seeds of a large number of uncultivated plants often carried by the wind or compost made of cow dung to germinate easily. It is quite interesting to mention here that there is a global estimate of 15 to 30 species of crops which feed the world, but two experts (Christine and Robert Prescott – Allen) have already identified 103 species providing 'the largest part of the world's plant food supply by volume.'[26]

Finally, all attempts should be made to preserve the uncultivated plants so that in combination with the cultivated ones the requirements of man as well as animals are met without taking a collision course harming the very purpose of meeting food-demand as a whole.

V. Survival of the Traditional Rural Agriculture as the most Important Tool for Achieving Food Sovereignty

Land grab, as mentioned earlier, starting around 2008 and according to Oxfam report foreign speculators have taken control of 227 million hectares of land in different regions of Africa, Asia and South America through clandestine deals between investors and politicians forcibly taking away the best tillable lands from the hands of poor farmers leaving them unprotected with loss of land, the only means of survival, in fact taking away the only resource of the poorest people in the world. The main countries getting involved in grabbing land elsewhere were China, South Korea and Saudi Arabia in order to ensure their food security in future in which the corporate giants like Monsanto, Dupont, Syngenta, Bayer, BASFand Dow participated in a different way – genetic engineering to change the character of existing crops. But as Wayne Roberts (2013, New Edition) puts it, it is again the peasant farmers who led the way to real innovation in agriculture through patience making a giant step in the name of 'Food Sovereignty'. It is highly interesting to know that:

> Perhaps the greatest period of genetic diversification in agriculture during the short span of human history is known as the 'Columbian Exchange' when domesticated plants from the Americas – corn (maize), potatoes, cassava, tomatoes, cocoa to name the best known – spread around the world within a hundred years, thanks to peasant innovators. The same was true for about 50 plants from Africa – coffee, cowpeas, black eyed peas, okra and sorghum, as well as rice varieties hidden in the hair of women as they crossed the ocean on slave ships,

when Africans were slaved to labour on plantations in the Americas.

Peasants and small farmers can perform the same miracles again, says Mooney. They work with 7,000 plant species, 84,000 breeds of land animals and 22,000 maritime species. They raise the food eaten by two-thirds of the world's population, using only 25 per cent of the planet's agricultural land. History shows they can do it amazingly, he says, 'if they can keep diversity, we will survive global climate change. If not, we won't. It is as simple as that'. Not a bad reason to think that 'food sovereignty is an idea for the future.'[27]

According to Jules Pretty, sustainable-agriculture expert from Essex University in England claiming that 'we can feed and nourish the world', and making an investigation of 208 'innovative farm projects' run by 9 million farmers in 28 countries found that those 'impoverished farmers' have raised, or brought land under their stewardship by '28,000 per cent, from 1,00,000 to 2,90,00,000 hectares'. What Pretty further discovered was that their efforts built up a system of interconnected processes providing 'health, environmental and economic dividend'. Such as, fish nurtured eat insects which indeed otherwise spread malaria while fish excrement enrich the soil in return providing a protein-rich meal to the farmer. All these findings urged the FAO to highlight the usefulness of these methods strongly recommending the adoption of the organic practices, making use of locally available 'low-cost and natural assets', such as animal manure, 'the strongest feature of organic agriculture', and according to the study of 293 cases, Ivette Perfecto from the University of Michigan similarly concluded that 'organic methods can yield over 2,641 calories per person per day'.

The great success of organic agriculture is further shown with examples citing a few cases of achieving *world-record* production in the rural areas of India. For example, a 'smallholder farmer called Sumant Kumar living in a tiny village of Darveshpur in Bihar, India's poorest state, produced 22.4 tonnes of rice on one hectare of land applying 'composted manure to soil and planting rice-seedlings 25 centimetres apart in a dryland. This method, known as system of Rice Intensification (SRI) was developed by a French Priest, Henri de Laulanie, working in Madagascar. The process increases yields and reducing water use cuts down 'methane production when rice grows under water'. Besides Sumant Kumar, 'his friend Nitish' broke the *world record* for potato yields and another farmer from a nearby village did the same for wheat. It can hardly be denied now that:

> People from Global South play leading roles in the food movement, thereby turning traditional colonial-era North-South relations right-side up. People in the Global South formed La Via Campesina, the largest general food and agriculture organisation in the world. In addition, Global South practices inspired many innovations coming out of the global food movement, community kitchens, community baked ovens, farmers' markets, seed exchanges, street vending, food trucks, urban agriculture, and gardening in vacant, public spaces. Critics of food movement in the Global North commonly refer to such practices as elitist, which may amuse those who developed the survival techniques of impoverished people, who knew to compensate for lack of money with richness of informal community.[29]

The growing popularity of organic food is a great support of organic agriculture practised by the world's poor in support of the food sovereignty movement. This is clearly evidenced by statistics provided by Wayne Roberts in Table 3.1.

Table 3.1: Organic Food Sales and Production Since 1990

1990	$1 billion in US food and non-food sales
2007	$1.7 billion in US food sales
2008	$3.2 billion in US food sales
2009	37.2 million hectares in worldwide production
2010	54 per cent of organic sales in mainstream and mass-market outlets
2010	$59.1 billion in global sales
2011	17,281 certified US farmers, ranchers, and processors
2011	$29.2 billion is US food sales
2011	4.2 per cent of US food sales are organic
2012	$1.3 billion in Australian food sales
2012	$8.4 billion in German sales
2012	$4.7 billion in French sales

Source: Wayne Roberts, 2013, New Edition: p. 152 (Taken from Agricultural Marketing Resource Centre: Organic Trade Association).

On the whole, sustainable agricultural practices provide a firm basis for providing mostly an assured base for supplying food for mankind. Jules Pretty (2010) is of firm opinion that agricultural practices, which, in other words may be called resource-conserving practices play a crucial role in making a firm basis for food sovereignty. Some of the most significant progress towards sustainable agrosystems has been achieved in developing countries as revealed in the analysis of 286 projects in 57 countries through the initiative of government, non-government, and international organisations. These 286 projects involved a total of almost '12.6 million farmers on 37 million hectares of land' engaged in transition towards agricultural sustainability covering just over 3 per cent of the total cultivated land (1.136 million hectares) in developing countries with the 'largest number of farmers in wet-land rice-based systems, mainly in Asia' and the largest area was in 'mixed systems, mainly in South America'. Table 2.2 shows the result of the adoption of Agricultural Sustainability programme from 286 projects in 57 countries.

Table 3.2: Agricultural Sustainability Programme

FOA Farm System Category	Number of farmers adopting	Number of hectares under Sustainable Agriculture	Average % Increase in Crop Fields
Small holder irrigated	1,77,287	3,57,940	130
Wetland Rice	87,11,286	70,07,584	22
Small holder rainfed Humid	17,04,958	10,81,071	102
Small holder rainfed Highland	4,01,699	7,25,635	107
Small holder rainfed dry/cold	6,04,804	7,37,896	99
Dualistic mixed	5,37,311	2,68,46,750	77
Coastal artisanal	2,20,000	1,60,000	62
Urban-based and kitchen garden	2,07,479	36,147	146
All projects	1,25,64,774	3,99,52,903	79.2

Source: J. Pretty, 2010, "Can Ecological Agriculture Feed Nine Billion People?" in Fred Magdoff and Brian Tokar (Edition), 2010, *Agriculture and Food in Crisis: Conflict, Resistance and Renewal*. Monthly Review Press, New York: Table 16.1, p. 293.

As it appears, the practice is progressing quite fast involving more farmers and more land and sustaining the progress in this respect it is expected that the agroecological practices are quite capable of feeding the estimated 9 billion people of the world by mid-21st century.[30]

NOTES

***U.N. 2014.** *Revision of the World Urbanisation Prospects*. p.1.

UN 2014 Revision of he World Urbanisation Prospects, p. 1: 'Today, 54 per cent of the world's population lives in urban areas, a proportion that revision of the World Urbanisation, is expected to increase to 66 per cent by 2050. Projections show that urbanisation combined with the overall growth of the world's population could add another 2.5 billion people to urban populations by 2050, with close to 90 per cent of the population concentrated in Asia and Africa, according to a new United Nations Report launched today. The 2014 revision of the World Urbanisation Prospects by UN DESA's Population Division notes that the largest urban growth will take place in India, China and Nigeria. These three countries will account for 37 per cent of the projected growth of the world's urban population between 2014 and 2050. India is projected to add 404 million urban dwellers, China 292 million and Nigeria 212 million.The urban population of the world has grown from 746 million in 1990 to 3.9 billion in 2014. Asia, despite its lower level of urbanisation, is home to 53 per cent of the world's urban population, followed by Europe with 14 per cent and Latin America and Caribbean with 13 per cent. The world's urban population is expected to surpass six billion by 2045.

1. Fitzpatrick, Tony *Climate Change and Poverty: A New Agenda for Developed Nations*. Bristol: Policy Press, 2014, p. 126. (Taken from Lang et al. *Food Policy: Integrating Health, Environment and Society*. Oxford: Oxford University Press, 2009, p. 200).
2. *Ibid.*, p. 128.
3. The World Bank: The World Development Report 2003. *Sustainable Development in a Dynamic World: Transforming Institutions, Growth and Quality of Life*, p. 83.
4. The World Bank: The World Development Report 2008. *Agriculture and Development*, pp. 2-4.
5. *Ibid.*, p. 6.
6. Roberts, Wayne. *The No-Nonsense Guide to World Food*. U.K.: New Internationalist, 2013, p. 26.
7. *Ibid.*, p. 27.
8. Wittman, H., Demaris, A., Wiebe, N. "The Origin & Potential of Food Sovereignty", in *Food Sovereignty: Reconnecting Food, Nature and Community*, ed. H. Wittman, A. Demaris, N. Wiebe, U.K.: Pambazuka Press, 2011, p. 2.
9. Altiere, Miguel A. "Agroecology, Small Farms and Food Sovereignty", in *Agriculture and Food Crisis: Conflict, Resistance, and Renewal* ed. F. Magdoff, B. Tokar. New York: Monthly Review Press, 2010, pp. 253-254.
11. *Ibid.*, p. 255.

12. Roberts, Wayne. *The No-Nonsense Guide to World Food.* U.K.: New Internationalist, 2013, pp. 90-91.
13. *Ibid.,* pp. 92-93.
14. Holt-Giménez, Eric. "From Food Crisis to Food Sovereignty", in *Agriculture and Food Crisis: Conflict, Resistance, and Renewal,* ed. F. Magdoff, B. Tokar. New York: Monthly Review Press, 2010, pp. 208-209.
15. *Ibid.,* pp. 210-211.
16. *Ibid.,* pp. 210-212.
17. *Ibid.,* p. 215.
18. *Ibid.,* pp. 215-217.
19. Roberts, Wayne. *The No-Nonsense Guide to World Food.* U.K.: New Internationalist, 2013, p. 96.
20. *Ibid.,* pp. 96-100.
21. *Ibid.,* pp. 100-103.
22. Smil, Vaclav. *Feeding the World: A Challenge for the Twenty First Century.* Cambridge, Massachusetts: The MIT Press, 2002, pp. 193-194.
23. *Ibid.,* pp. 194-195.
24. Mazhar, F., Buckles, D., Satheesh, P.V., Akhtar, F. *Food Sovereignty and Uncultivated Biodiversity in South Asia: Essays on the Poverty of Food Policy and the Wealth of the Social Landscape*, New Delhi: Academic Foundation, 2007, p. 2.
25. *Ibid.,* pp. 4, 5, 11-12.
26. *Ibid.,* pp. 15, 18-19.
27. Roberts, Wayne. *The No-Nonsense Guide to World Food.* U.K.: New Internationalist, 2013, p. 114.
28. *Ibid.,* pp. 142-143.
29. *Ibid.,* p. 146.
30. Pretty, Jules. "Can Ecological Agriculture Feed Nine Billion People?" in *Agriculture and Food Crisis: Conflict, Resistance, and Renewal,* ed. F. Magdoff, B. Tokar. New York: Monthly Review Press, 2010, p. 297.

4
Earth's Resources and Smallholder Farming Practices
A Study in Compatibility

I. Preamble

Talking about agriculture, people often forget about what type of agriculture is referred to since its practice widely differs from people to people as well as from area to area, meaning the geographical setting. Thus, the practice in America or Europe differs widely from Africa and Asia in every aspect. For example, the crops grown in America or Europe, or even in Australia are largely not for direct consumption of people either who grow them, or for sale in the local market. Instead, there are large business concerns who buy them in bulk to be sold to another set of large business houses who buy in bulk to be converted into processed or semi-processed food to be directly consumed by urban people in general who want to be saved the inconvenience of cooking food and there are a growing number of supermarkets catering to the needs of daily food for consumption already properly or partly cooked. In other words, the supermarkets making a link in the chain connecting growers and buyers/consumers have come to be a part of this agricultural process, mostly controlled by big industrial set-ups, giving agricultural practices an altogether new character, an industry by itself where even the producers have no right to choose the type and amount of the crops to produce because they no longer remain landowners, working under the protective wings of government, the final producers working in close collaboration with giant corporate houses. Further, these corporate houses decide if the crops they buy in huge bulk are going to be used directly for human consumption or put to use for other purposes, producing more meat for rich buyers or converted into biofuel/ biodiesel for transport.

However, our major concern here is about smallholder rural farmers concentrated mostly in the Global South who still hold major control on food production for their own consumption as well as consumption for people around the world. However, the situation is not as simple as that. The smallholder peasants cannot often work independently as in the majority of cases they are land-less farmers or own a very small acreage which is often not of economic-size. Besides, the crops they produce, say rice, potatoes or any other vegetables etc. find market nearby or distant urban areas through the hands of intermediaries making a larger share of the profit as the farmers neither have direct access to the market lacking transport, storage or in most of the cases they are committed to the intermediaries before even the production process begins taking money in advance. Hence, the entire process from production to selling the produce are controlled by the intermediaries while the farmers themselves in majority of cases are in the clutches of rural moneylenders realising high rates of interest – in the absence of bank loans – at the beginning of the season when the farmers need money for buying seeds and other accessories (including sometimes hired extra labour or quite often bullocks/animals to draw the ploughs). The moneylenders, in the absence of bank or cooperative loans (a practice in India), take full advantage of grossly inflating the calculations with illiterate farmers and since the farmer quite often cannot clear the loan after the harvest, the loan-amount along with high interest goes on multiplying, keeping the farmer perpetually indebted for generations, giving rise to a system of slavery from which he and his family never get released without proper intervention from government (which happened in recent years in West Bengal, India ending the rule of slavery for good).

But however brutally mauled, the rural farmer had been in the unscrupulous hands of a system over a long period, the rural farmers did not take leave of their job they had been doing for generations and as mentioned in Chapter 2, the number of rural peasant farmers has more or less remained the same and their contribution of food to the world population still plays a significant role.

Now, we need to find out how far the Earth's existing resources are conducive to sustain the practice of rural peasant farmers since it is this type of agriculture which unbound by restrictions of industrial houses are free as the only means of feeding the world in general and its growing population engaged in non-food productive activities besides facing the increasing challenge of food-crops getting converted into bio-fuel or high-protein food like meat becoming more affordable for people with rising income, even for a section of the Global South.

II. Earth's Resources and Ecosystems

In assessing Earth's resources, they are divided into eight Ecosystem zones primarily based on their physical characteristics:

1. Agroecosystems
2. Coastal Ecosystems
3. Forest Ecosystems
4. Freshwater Ecosystems
5. Grassland Ecosystems
6. Mountain Ecosystems
7. Polar Ecosystems
8. Urban Ecosystems

The basic difference in the physical characteristics of the earth, such as temperature regime changing towards north and south of the equator, the wind systems as controlled by sun and other factors, the landscapes with their widely changing topography ranging from marshy land, low and high plateaus, hills and mountains etc., ultimately have defined the character of ecosystems which emerged with great variations in resources defining human actions.

Natural resources naturally vary very widely between the ecosystems responding differently again in supporting human activities. This is very clearly reflected in the Agroecosystem which as its name claims, providing 'overwhelming majority of crops, livestock feed and livestock' amounting to '95 per cent of all animal and plant protein' in 1997, as mentioned in United Nations World Resources 2000-2001 besides fulfilling a large part of the requirement of fibres of different types of human use.

Thus, the agroecosystem has turned out to be the most human-centric ecosystem responding to human use of its resources in defining corresponding human activities among which agriculture takes a very dominating role offering great scope for different types of agricultural activities to flourish. The associated expansion of agricultural activities naturally has involved use of more land gradually coming into clash with the use of land for other purposes like 'industrial, commercial or residential (urban extension) purposes.'

On the other hand, intensification of production involving higher yield from the same land lacking any further scope for expansion has given rise to a system of production particularly in Asia, with multiple crops grown from the same land with the aid of irrigation and using new varieties of crops with shorter growing seasons. At the same time, the system of continuous cultivation on 'high-quality non-irrigated land has been made practicable by abandoning or shortening the fallow period'. As a matter-of-fact intensification of the productive system has become more popular in areas close to urban areas providing a ready and very responsive market for fresh products provided by the rural farmers consisting of 'high value perishables such as dairy products and vegetables'.

However, the backlash of this highly intensive agricultural system run with great economic success of the poor peasant farmers started gradually being felt in the slow deterioration of the physical conditions affecting the productive system in the long run unless more cautionary steps are taken in time to check their process of environmental deterioration. 'These stresses include increased erosion, soil nutrient depletion, salinisation and waterlogging of soils, and reduction of genetic diversity among major crops'.[1]

Agroecosystems cover about 36 million hectares of Earth's land surface (excluding Greenland and Antarctica), but it does not appear in one continuous stretch of cropland. Instead, according to the Food and Agriculture Organisation (FAO), '69 per cent of agroecosystems consist of permanent pastures' and the remainder under crops. Further, according to the FAO, there is wide difference in the combination of land under crops or pasture. Croplands are predominant in South Asia comprising 92 per cent of agroecosystems and 84 per cent in Southeast Asia. 'In India 94 per cent of the agroecosystem area is covered by crops of which 91 per cent of the land is engaged in the production of crops like wheat, rice, maize and soyabeans while the remaining land produces tea, coffee, sugarcane, and most fruits. In contrast, pasturelands predominate in agroecosystems in other areas, such as it is 89 per cent in Oceania, 83 per cent in Sub-Saharan Africa, 82 per cent in South America and 80 per cent in East Asia'.[2]

According to the FAO (2000) estimates, total agricultural area expanded slowly between 1966 and 1996, from 4.5 billion hectares to 4.92 billion hectares, an 8 per cent increase. But conversely agricultural area was being taken away from agricultural production for other uses in the United States, Western Europe and Oceania amounting to a total of 49 million hectares over a period of three decades besides in Eastern Europe exhibiting the same trend. In contrast, however, agricultural land in South Asia has remained constant for more than 20 years at about 223 million hectares while both in China and Brazil this has increased by nearly 0.8 per cent per year between 1986 and 1996 besides a slightly larger increase in West Asia recording 1.38 per cent per year.

There have also been quite remarkable changes in intensification in agricultural operations in geographical ecosystems. This particularly involves growth in irrigated area over the past three decades, between 1966 and 1998 from 153 million hectares to 271 million hectares comprising globally only 5.5 per cent of the total agricultural land, or 17.5 per cent of the cropland though in some areas, like in China and India together represent 41 per cent of the global irrigated area while the United States and Latin America together contain 12.5 per cent of the total. In contrast, however, it is a paltry 3 per cent of the total global area for both Sub-Saharan Africa and Australia.

Intensification of agriculture is most prominent in Asia where all of the cropland is harvested each year sometimes 'two to three times a season' using 'irrigation, new types of quick-growing seeds, and fertilizers and abandoning the practice of leaving the land fallow'. The increases in land productivity have naturally been of great economic value to the farmers of Asia, the majority of whom had always suffered from malnutrition. It is estimated by FAO, that 790 million people do not have 'adequate access to food', making 33 per cent of the population in Sub-Saharan Africa, 31 per cent in the Caribbean and 23 per cent in South Asia 'undernourished'.[3]

However, the overall importance of agriculture in human life may be gauged from the fact that 'the food, fibre and animal feed that is provided by world's agroecosystems amount to about $1.3 trillion per year' and agriculture employs 46 per cent of the total global labour force while almost half of the total population living in rural areas depend on agriculture as their livelihood. Another side of agroecosystem's role is carbon storage 'mostly in the soil amounting to about 26-28 per cent of total terrestrial carbon made possible through improved nutrient management, reduced soil erosion and widely adopted use of no, or minimum tillage.'[4]

In sharp contrast to agroecosystems, the coastal ecosystems comprising the 'intertidal and subtidal areas above the continental shelf (to a depth of 200 m) and adjacent land up to 100 km inland from the coast' represent a 'much more diverse array of habitats', than any other ecosystem which also includes coral reefs, mangroves, tidal wetlands, estuaries, peat swamps' providing altogether a much larger variety of goods and services than any other ecosystem. The coastal ecosystems are inhabited by almost 39 per cent of the global population. Numbers living within 100 km of the coast increased from 2 billion in 1990 to 2.2 billion in 1995 – the most populated ecosystem in the world and the one most physically affected by inhabitants making changes in the environment to make it suitable to meet needs of 'shelter, subsidence, commerce and recreation'. As a result, almost '29 per cent of all lands within 100 km of a coastline has been altered significantly to suit human purpose, of which

19 per cent are highly altered for practising agriculture and urbanisation while another 10 per cent is partly altered interspersed with agriculture and natural vegetation, leaving almost 71 per cent of the land under this coastal ecosystem to its natural self.[5]

Unlike the other ecosystems, the coastal ecosystems are not only the most varied in their physical characteristics but also include quite a few major regions of both extreme physical and economic importance. Foremost among them are the mangroves and coral reefs playing quite a critical role in developing a different set of human occupations.

Among them the mangroves take special place in the Earth's resources. According to this Report, mangroves 'line 8 per cent of the world's coasts' and nearly a quarter of them occur in the World's tropical coastlines, covering a 'shore-area of about 1,81,000 km^2', thriving best in sheltered coastal areas such as river estuaries. Acting as spawning grounds for a wide range of fish species they play a crucial role to productivity of tropical fisheries besides providing local inhabitants with timber and fuel stabilising the coastline. However, uncontrolled human action in cutting timber and fuel as well as clearance for agricultural expansion over the years has already caused heavy damage to the mangrove forests, leading to their permanent loss ranging from 4 to as much as 84 per cent, mainly in the recent decades. An idea about the coverage and loss in extent of the mangroves in selected countries may be obtained from Table 4.1 to ascertain its likely effect on local economy.

In brief, of the 22 selected cases spread over four continents, the loss is at a highly critical stage (with loss of mangrove) ranging between 50 to 84 per cent, in 14 out of a total of 22 regions spread over four continents exceeding a loss of 70 per cent and over in four out of eight cases in Asia.

Coral reefs constitute another very interesting feature occurring in shallow tropical waters with minimum silt content occupying nearly 2,55,000 square kilometres of the world's surface.They support nearly 1 million species of plants and animals along with rich biodiversity supporting small-scale fishing and protecting the coastline from storm damage. More than two-thirds of coral reefs are concentrated on the coasts of the Western Pacific and Southeast Asia where the Indo-Pacific Region contains 91.4 per cent or 2,33,000 square kilometres of coral reefs and, in general, they are extremely vulnerable to human activities like 'land reclamation, coastal development or coral mining' besides siltation, pollution, rise of sea level and temperature.[6]

Table 4.1: Mangrove Area in Selected Countries

Region and Country	Current Extent (km²)	Approximate Loss (%)	Period
Africa			
Angola	1,100	50	Original Extent to 1980s
Cote d'Ivoire	640	60	Original Extent to 1980s
Gabon	1,150	50	Original Extent to 1980s
Guinea-Bissau	3,150	70	Original Extent to 1980s
Kenya	610	4	1971-1988
Tanzania	2,120	60	Original Extent to 1980s
Latin America and the Caribbean			
Costa Rica	413	-6	1983-1990
El Salvador	415	8	1983-1990
Guatemala	161	31	1960s-1990
Jamaica	106	30	Original Extent to 1990s
Mexico	5,315	65	1970s-1990
Panama	1,581	67	1983-1990
Peru	51	25	1982-1992
Asia			
Brunei	200	20	Original Extent to 1986s
Indonesia	24,237	55	Original Extent to 1980s
Malaysia	2,327	74	Original Extent to 1992-1993
Myanmar	4,219	75	Original Extent to 1992-1993
Pakistan	1,540	78	Original Extent to 1980s
Philippines	1,490	67	1918-1980s
Thailand	1,946	84	Original Extent to 1993s

contd...

Region and Country	Current Extent (km^2)	Approximate Loss (%)	Period
Vietnam	2,525	37	Original Extent to 1993
Oceania			
Papua and New Guinea	4.627	8	Original Extent to 1992-1993

Source: The World Bank: World Resource 2000-2001, People and Ecosystems: The Fraying Web of Life, p. 74.

Fishing is the most important source of earning in the coastal ecosystems providing jobs to nearly 29 million people, some 95 per cent of whom are in developing countries where poor fishing-folk find it as the major source of livelihood. However, globally the value of 'fishery exports in 1996 amounted US$52.5 billion, 11 per cent of the value of agricultural exports that year'.

The most important ecosystem from human point of view is the Forest Ecosystem. There are two estimates about total forest coverage, one of them claiming it to be 29 million square kilometres in 1993, covering about 22 per cent of the global land area (excluding Antarctica and Greenland), while the other claims it to be 34.5 million square kilometres in 1995, taking up 27 per cent of the global land area (UN World Resources 2000-2001).

But these calculations are far from static as the rates of forest depletion, more in the recent times, are very high and according to FAO's estimate 'roughly 20 per cent of the global forest cover had been reduced by early 1980s, while a study by the World Bank in 1997 put it at 50 per cent though one FAO study claims that the rate of forest loss in developing countries decreased by 11 per cent and forest cover decreased by 0.2 million km^2 (2.7 per cent) in industrialised countries between 1980 and 1995. The causes of forest loss differ widely from being expansion of agriculture in Africa according to FAO's assessment in 1997 to in Latin America being 'due more to large-scale cattle-ranching besides government-planned settlement scheme and hydroelectric reservoirs.' FAO found in Asia the pressure of subsistence agriculture, naturally made by rural farmers, and economic development schemes responsible for clearing forested areas.

Forest fires, on the other hand, plays a very dominant role in reducing forests causing inestimable losses of flora and fauna with very upsetting effects on national economy. Regrettably, forest fires may be man-made as well as natural, when fires are 'intentionally' made for 'timber harvesting', land conversion or shifting agriculture! They are greater in number than those naturally caused ones. On the whole biomass burning reshaping forest ecosystems amount to as much as '90 per cent of the total biomass burning including the savannas'. Both social and economic costs of forest fires are enormous. For example, an 'estimated 20 million people' suffered from 'respiratory problems in Southeast Asia' in 1999 when economic damages were estimated at $4.4 billion. Forest fires are not limited to tropical areas. Even the boreal forests in the United States and Canada suffer from forest fires, the area of fires having doubled during the three decades up to 1999.

In fact, the role for forest ecosystem is equally affected by human action, dangerously undermining its role in the service of man. One of the major instances cited is that since 'wood fuels will remain of prime importance in developing world' necessary steps need to be taken in this direction. At the same time:

> Forest biodiversity is a good in its own right. Diverse species found only in forest habitats are sources of new pharmaceuticals, genetic resources, and non-timber forest product such as resins, fruits, vines, mushrooms, and livestock fodder. Even more important, all other forest goods and services depend to some extent on the diversity of forest species.[7]

In fact, forests take an important role in biodiversity conservation. Of the 136 identified biodiversity regions, two-thirds are located in the Forest Ecosystems. Of the 234 centres of plant diversity identified worldwide, more than 70 per cent are found in the forests while of 596 key area having threatened species, more than 70 per cent are located in forests. All these are very strong indications for the loss of both tree-types and animal life in the biodiversities located in the forest ecosystems threatened by human action in the wrong direction.

Similarly, wrong human action is further playing havoc in diminishing one of the most important roles of forest ecosystems taking a central role in the global carbon cycle. 'Trees capture carbon from the atmosphere as they grow and store it in their tissues' and with a great biomass, global forests turn into terrestrial '*sinks*' of carbon, storing almost '39 per cent of the total carbon'. But deforestation and simultaneous other human activities degrading the forests has caused 20 per cent of carbon emission.

In comparison to these three ecosystems – the Agroecosystems, the Coastal and Forest Ecosystems – other ecosystems play a less significant role in human life except for the Grassland Ecosystem providing home to about 938 million people in the world, occurring both in humid as well as arid zones. Its role in shaping human history cannot be ignored since the 'ancestors of nearly all the major cereal crops, like wheat, rice, barley, rye, millet and sorghum originated in the grass lands and though agroecosystems have 'replaced many grasslands', yet they provide even now 'genetic resources for improving food crops'. Their four major roles are: food production, biodiversity maintenance, carbon storage and tourism.

Geographically, grasslands maintain quite a wide sweep, encompassing 'non-woody grasslands, savannas, woodlands, shrublands and tundra', thus, appearing on every continent, covering nearly 41 per cent of Earth's land area (excluding Antarctica and Greenland), and turning out to be the most extensive of all ecosystems nurturing human life. Grassland ecosystems cover 52.5 million km^2, while covering more than 50 per cent of land area in 40 countries in 20 of which – most of them in Africa – grasslands' share rises to over 70 per cent of the total area.

However, scientists are of firm opinion that a large portion of the grasslands have been converted to agriculture and in this regard the 'Tall-grass Prairies in North America shows the greatest change' where 71 per cent of this region is covered by croplands and 19 per cent by urban areas whereas in contrast the situation in Asia, Africa and Australia is quite different where 50 per cent of the area keeps the grassland cover, less than 20 per cent of the grassland has been taken over by cropland and less than 2 per cent for urban or similar uses. In South America, on the other hand only '21 per of grassland remains undisturbed.'[8]

The role of grasslands is central, or pro-genital to food production having provided the most leading grass-crops as mentioned above and most extensively having been transformed into agricultural lands. But today where not intruded by crop production, the grasslands have turned out to be the livestock meat producer, providing a major livelihood to rural population, especially in Africa. Grasslands' overall contribution to beef production amounted to over 5 per cent of global total of 54 million tons in 1998 while 'mutton and goat output increased to over 26 per cent, or nearly 11 million tons' (UN Environment Programme).

Grassland fires are another characteristic feature of this ecosystem. In many African countries deliberate burning of grassland, ranging from 25-50 per cent of arid Sudan zone to as much as 60-80 per cent in the humid Guinea zone takes place, which is very harmful for the release of carbon to the atmosphere. However, grasslands store about 23 per cent of all carbon stored in terrestrial ecosystems mostly in the soil though it is getting reduced through conversion of grasslands into other uses. Among the remaining ecosystems, Freshwater, Mountain, Polar and Urban, it is only the mountain ecosystem which plays some important role among Earth's resources in contributing to human food-supply. The mountain ecosystem, representing areas above 3000 metres cover about 5 per cent of the world's terrestrial surface, but 27 per cent considering land above 1000 metres.

In fact, the actual coverage of the Earth's land surface by each ecosystem is far from clear as there is too much of intermixing, one grading into another, or rather overlap, such as agroecosystems overlap with forest and grassland. For example, according to FAO '69 per cent of agroecosystems consist of permanent pasture' and the position varies widely from region to region, permanent pastures representing '89 per cent of the agroecosystems are in Oceania in contrast to croplands covering 92 per cent land of the agroecosystems in South Asia, as mentioned earlier, and it is 92 per cent in India'. This overlapping is the case with each of the three other ecosystems – Coastal, Forest and Grassland – providing Earth's resources primarily for human uses. For instance, the Coastline Ecosystem is in constant change, with 29 per cent of its land altered due to human use. At the same time, the Grassland Ecosystem has experienced drastic changes with extension of croplands and other land-uses converting 40 per cent of the total area. As a result, these different ecosystems with their immense resources overlapping with each other at many instances involving the major part of the Earth's land surface make one thing very clear which is the use of the resources for the purpose of providing greater support to activities beneficial to mankind's survival. Our major aim in this respect is to identify the character of overlapping which appears very clearly to be more expansion of agriculture than other human activities connected to urbanisation and industrialisation.

Covering about 4 per cent of the world's land surface (urban) they are far from likely to disturb agricultural areas beyond the suburbs for the reason that they will not disturb the food-channel connecting them with the surrounding rural areas. Extensive channels of food-imports developed all over the world have no alternative for fresh supplies from nearby rural agricultural-zone operated by rural farmers compared with the choice of industrial agriculture taking place elsewhere.

III. Role of Agriculture in Facing Worsening Food Situation

Agriculture has always had as its main aim feeding people without questioning the status of the fed. Yet about '820 million people' do not have adequate food and suffer malnutrition while 160 million children are seriously underfed. As has already been stressed in Chapter 2, it is the rural agriculturists, or in other words, the small peasant farmers in the Global South who feed the world, and not the industrial agriculture run by corporate houses. In another 30 years from now, by 2050, an additional '2 to 3 billion' people globally have to be fed besides the need to 'improve the diets of 2.5 to 3 million people living on less than $1 a day'. To make that happen, some very important steps have to be taken requiring sincere and honest participation of local governments as well as lawmakers of the countries involved.

> That will require tilting institutional rules to move assets into the hands of small-landholders, halting nutrient mining, reducing soil erosion, and adopting agricultural practices that restore soil fertility. It will also require sharing rural land and water to expanding urban population and meet environmental needs.[9]

This is exactly what led to the idea of Food Sovereignty which emerged as the only alternative to capital-intensive corporate-led agriculture, the former called La Via Campesina, referred to earlier in the previous chapter. As described earlier, food sovereignty is broadly defined as 'the right of nations and people to control their own food system'. It proposed that: 'the control of productive resources (land, water, seeds and natural resources)[be placed] in the hands of those who produce food.'[10]

However, to make food a fundamental right, an unavoidable entitlement like freedom for every human being is to find out if there is enough land for meeting the requirement of a growing population coupled with more urban-ward movement in future when, as mentioned before, the global urban population will be 66 per cent of the total population. Incidentally, agriculture all over the world has made serious inroads into other land uses, such as 'one-fifth of the tropical forests have been cleared giving way to agricultural expansion since 1960 and thus destroying the biodiversity leaving it confined hardly to 1.4 per cent of the world's land surface with concomitant increase in CO_2 emissions along with great environmental costs when forests removed had tremendous effect not only on holding back carbon as a sink but also filtration of water through its soil for purification with loss of water-flow.

It has been repeatedly mentioned, as in 2007-08, there is no dearth of food in the world, but lack of purchasing power for 70 per cent of the rural population. The essential steps that are needed to be taken are 'increasing their ability to produce more to consume and food to sell in the markets'. But growing income of a section of the people in the Global South will naturally make larger demand for food in the long run and rural people would require more land and water to produce more and keep the price under control. All this is possible with further expansion in agricultural land and more water available for irrigation where it is needed.

The information provided in The World Bank Report 2003, points out the limitation of availability of additional land for agricultural expansion in the developing world. According to their calculations, severe degradation from water erosion particularly in Southeast Asia and Central America 'soil nutrient mining' particularly in Africa, and salinisation in areas under large irrigation systems, such as in India, have jointly caused an annual loss of agricultural land of about 0.5 per cent globally. As a result, almost as much land has been lost in this process totalling '303 million hectares as is now used – 307 million hectares in Africa and Latin America.'

Conversely, 'these two regions, Africa and Latin America', now possess 'extremely large reserves of tillable land – 700 million hectares and 680 million hectares respectability, 45 per cent of which lies in forests.' In sharp contrast hardly any additional tillable land exists in East Asia and the Pacific, or in Middle East and North Africa while in South Asia already farming is done on soils not 'sustainable for cultivation'.[11] However, most of the land available in Africa and Latin America suffer from many limitations to bring them under cultivation and as such the prospect of agricultural expansion is extremely limited.

The second important physical factor for agricultural expansion is water availability. According to the International Water Management Institute (IWMI) use of potentially available water is very poor in Africa where only South Africa has exploited more than 20 per cent of it while in Latin America apart from Cuba using 48 per cent and Mexico 27 per cent much of future irrigation prospects in these regions looks quite limited. As a matter of fact, the contrasting physical situation in Africa has kept irrigation confined to only 1-3 per cent of its agriculture under irrigation while expansion is kept limited by natural factors like '60 per cent of irrigation potential is in humid region where rain-fed agriculture at best could be supplemented by irrigation if at all needed'. In contrast, according to the same source, IWMI, 'China is the most water-stressed country in East Asia using 44 per cent of the useable water' while projected 60 per cent use by 2020 is considered to exceed 'environmental capacity'(IWMI).

It has been pointed out in The World Bank Report 2003 that the social responsibility of the rural sector is going to be enormous in the coming 30 years or so, that is by 2050, when it will be burdened with the responsibility of feeding the largest share of the projected world population of 9 billion. According to this Report, the first major step towards enabling rural farmers to meet this great task depends largely on breaking the poverty-trap by training in non-farming employment and introduction of improved techniques of 'soil fertility replenishment' developed by the International Centre for Research in Agroforestry and already adopted by tens of thousands of farm families in Kenya, Malawi, Mozambique, Tanzania, Uganda, Zambia and Zimbabwe' using 'various combinations of fallow, phosphorous and biomass transfers'. One of the very important steps which needs to be taken for family farmers is 'reduced agricultural taxation' as done in major Southeast Asian economies, like Indonesia, Malaysia, Thailand, Taiwan and the Republic of Korea. In this connection, China has gone further giving most agricultural land to farming families though it is a very 'small holding' for each (and sharply increased the price of the agricultural produce acting as a great inducement and recording a recognisable reduction in rural poverty).'[12] Finally,

> in poor developing countries with large agricultural sections, growth led by the agricultural sector has a powerful effect in pulling people out of poverty, especially when incomes and assets of the rural sector are somewhat equal. Small holders with assets develop voice and become political players. History has shown that this generates an inclusive development path that helps countries face later challenges. But getting the assets into the hands of small holders requires good land and water policies. These policies also enable poor people to get access to opportunities for building their human and social capital.[13]

According to The World Bank Report 2008, of '3 billion global rural population' about 2.5 million remains engaged in agriculture with 1.5 million of them living in small holder households making poverty overwhelmingly rural. Growth both in 'agriculture and rural non-farm economy is needed to reduce rural poverty'.

In this perspective agricultural growth has been quite impressive, expanding globally by an average of 2 per cent a year and exceeding the rate of population growth of 1.6 per cent per year between 1980 and 2004, and in this respect developing countries achieved a faster rate of growth – 2.6 per cent a year – than the industrial countries (0.9 per cent a year during the same period: 1980-2004). The development in growth in Asia and the developing world is contributed by 'productivity gains rather than agricultural expansion, but based primarily on 'widespread use for irrigation, improved crop varieties, and fertilizers'. Irrigation has played a great role in Asian agriculture going back in certain cases to millennia – such as in Harappan times (2500 B.C.-1500 B.C.) and today '39 per cent of the crop in South Asia, 29 per cent in East Asia and the Pacific, but only 4 per cent in Sub-Saharan Africa is irrigated'. Growth in crop production in the developing world has also been followed by increase in meat production having doubled in fifteen years along with poultry by 7 per cent annually.

Apart from technological aids, development in infrastructure as well as investment in human capital through education, health and nutrition have provided further incentive towards increase in agricultural productivity. Such an investment in the construction of rural roads resulted in high dividend in agricultural production to the tune of 25 per cent in the 1970s.[14] Progress in agricultural growth has been maintained well in Asia, particularly in China recording '3.5 per cent annually', while it had been 3.1 per cent in Malaysia followed by 2.4 per cent in Vietnam. But this has been highly variable among Sub-Saharan countries. Apart from Nigeria, Mozambique, Sudan, and South Africa which had agricultural growth above 2 per cent a year over the period of last 25 years, others had failed in doing it.

IV. Agricultural Growth through Diversification

As pointed out in The World Bank Report 2008 on 'Agriculture for Development', for 60 per cent of rural agricultural producers there are opportunities for further development through diversifying to higher-value products. Looking at the fast change in dietary habits of people particularly with rising income, agricultural producers should turn to the production of fruits, vegetables, vegetable oils and livestock products enjoying a growing market world-over. Already, a horticultural revolution is on its way with fruits and vegetables' markets growing in developing countries 'with production increasing by 3.6 per cent a year for fruits and 5.5. per cent for vegetables between 1980-2004. During this period, China produced 58 per cent of the global horticultural production, 38 per cent came from all other developing countries while the developed countries contributed only 4 per cent. India also forged ahead with the production of fruits as an important part of its agricultural crops since the 1990s. Diversification has created many jobs, particularly suitable for women in activities like 'processing, packaging and marketing'.

Another phase of agricultural diversification is represented by 'livestock and aquacultural revolution' taking a very important role transforming urbanised countries of Asia and Latin America 'driven by rising demands for poultry, pork, fish, eggs' along with beef and milk, demands of the latter having almost doubled in India between 'early 1980s to late 1990s'.

In the case of India, the livestock production has, meanwhile, undergone a sea-change from extensive (grazing) to intensive (stall-fed), rearing of poultry, pig and cows making a greater demand for food grains, including oil seeds using almost 28 per cent of the grain as feed of the livestock.[15]

At the same time aquaculture forms another most important part of agricultural diversification programme, constituting the 'world's fastest growing food production sector, increasing at an annual average rate of 10 per cent since the mid-1980s', representing over 30 per cent of total food-fish production, more than 90 per cent of it taking place in the developing countries, and among them China alone 'accounts for 67 per cent' of global-catch. Aquaculture provides scope for multifarious jobs (even cutting fish to the choice of the buyer provides a job). More than 12 million people in Asia earn a living from aquaculture while in Bangladesh as well as in Vietnam 'more than 50 per cent of workers employed in fish depots and processing plants are women' though working on very poor salaries.[16]

Finally, agricultural diversification happening mostly in the developing world, called Global South has great impact on international trade, making up the major share, '47 per cent of all developing-country export' consisting of livestock, fish, cut-flowers, and organic products' replacing traditional tropical products like, tea and cotton representing 21 per cent of total export. Interestingly, developing countries provided 43 per cent of the world trade in fruits and vegetables in 2004 (excluding bananas and citrus). Some countries like Brazil, Chile, China and Mexico have a dominating share in these exports, with Kenya becoming another partner.

V. A Bleak Future for Food? Major Constraints for Agricultural Sustainability

The major question is even if agriculture has so far been successful in meeting the global demand for food-barring the hungry billion getting deprived more for lack of purchasing power than lack of food (along with its distribution), the overall global situation is likely to get worse with increasing population as well as more people moving to urban areas all over the globe.

Assessing the future IFPRI (International Food Policy Research Institute) predicts that global cereal consumption growth is slowing down from 1.9 per cent annually in 1969-1999 to 1.3 per cent a year 2000-2030 along with slowing down of meat consumption from '2.9 per cent a year to 1.7 per cent annually', resulting from 'overall slowing of population growth to 1 per cent a year' in all developing countries though equally contrasted by 'medium to high levels of food consumption per capita' in some highly populous developing countries, such as in China – slowing down of consumption will naturally lead to slow down in production. Despite this, as IFPRI stresses there will be wide imbalances in imports in developing countries of Asia, Africa and Latin America together importing 265 million tons of cereals in 2030 increasing from 85 million tons in 2000, which reveals the actual reason of high dependence of Middle East and North Africa along with increases in imports in Asia and Sub-Saharan Africa.

In this backdrop, the major exporting countries are developed countries along with Brazil and Argentina while some countries in Europe and Central Asia are likely to become important exporters in future. But the entire situation is constrained by some physical barriers – without removing them no immediate solution is likely to be forthcoming. Such obstacles are presented by the availability of adequate supplies of land and water, growing climate variability and climate change, besides frustrating attitude of governments, in their reluctance to invest more in research, posing formidable challenges in meeting future food-demand.[17]

To supply food as the primary necessity has been increasingly interlinking with population growth exhausting the resources in some areas, such as in Asia, where the ceiling has been reached leaving no more scope for further expansion. However, this is not the case with many regions, such as in Latin America, Central Asia and even in Europe with 'lower population densities as well as declining agricultural population'. But whatever it is, conversion of more land for agricultural expansion becomes more and more uneconomic, as in Latin America, where it has to be done at the cost of removing tropical and subtropical forests and woodlands. Similarly, Sub-Saharan Africa provides scope for further expansion mainly in forest or grazing lands, but those areas are already infested with human and animal diseases removal of which requires huge investments.

Writing in 1996, Tim Dyson commented that:

> expansion of the area of cropland is rather limited ... and this is also true about the Middle East and South Asia where

> 'virtually all the estimated potential arable land has already been cropped. And the same is true throughout the Far East, where the major reserves of exploitable land were chiefly found in sparsely populated parts of Indonesia and Cambodia. The circumstances of Sub-Saharan Africa and Latin America are somewhat different. In these regions.... under one-fifth of potential arable land is currently cropped.... However, the main tracts of unexploited land are in locations like the Congo and Amazon Basins – which are quite far from the major regional concentrations.... Also, when such tropical forests are cleared for food production their soils often turn out to be not very productive over the long run.... However, looking at the world as a whole, it is clear that the vast majority of increased food demand will not be met by the simple conversion of new land to crops.... Instead, the overwhelming bulk of future food demand must be met through more productive use of existing cropland.[18]

On the other hand, according to Ehrlich *et al.* (1993), the expansion of food production had been achieved at a 'heavy cost – the depletion of a one-time inheritance of natural capital crucial to agriculture', or in other words, the loss of top-soil amounting to roughly 24 billion tons along with 'trillions of gallons of ground water' and 'species of other organisms' representing the ecosystem. According to this source, about 17 per cent of the earth's potentially productive surface is used for crops of which 593 million hectares were planted in grains in 1950 which increased to 720 million hectares in 1990, doubling the average yield under improved treatment.[19] Ehrlich *et al.* (1993) are very explicit about the constraints to increasing food production calling them bio-physical limits. There are, as they observe: (a) losses of farmland to other uses and population pressure and limits to the amount of suitable land that can be brought into production; (b) limits to fresh water supply gradually cutting down the proportion of irrigated land; (c) erosion and soil-degradation; (d) biological limits to yield; (e) diminishing returns from fertilizer-use; (f) problems associated with chemical control of pests; (g) declining genetic diversity of crops and their wild relatives; (h) declining yields from increased ultra-violet B radiation; (i) air pollutants affecting yield-rates; (j) effect of climate change on reducing production and rising sea level; (k) decline in ecosystem services.[20] On the other hand, IFPRI 2002 (International Food Policy Research Institute) observes that the 'changing face of farming practices' results from 'aging of the farm population, its feminisation and decreasing cost of capital relative to labour are putting the entire practice under great threat.'[21]

Among the major constraints affecting food production apart from land becoming scarce for further agricultural expansion, the productive capacity of land is often undermined by the deteriorating condition of land affected in many areas by salinisation as well as poor management causing degradation through the practice of intensification. This causes altogether an annual loss of 5 to 10 million hectares of agricultural land globally. Further, nutrient mining is a great problem in Sub-Saharan Africa, though rectifiable through better soil management and fertilizer use.

The second obstacle is presented by acute water scarcity. In developing countries where agriculture uses 85 per cent of fresh water producing 40 per cent of the total crops in value. The growing competition between agriculture and industrial – urban uses of water has worsened the situation further and according to the Comprehensive Assessment of Water Management, about '1.1 billion people' living in the river basins suffer from 'absolute water scarcity' while the situation is fast approaching for another '478 million living in river basins' and another '1.5 billion' suffer lack of access to water because of inadequate infrastructure and capital to tap the sources. On the other hand, 'unsustainable extraction of river or underground water is worsening the water-crisis in China, South Asia, Middle East and North Africa, overuse of ground water being '25 per cent in China, and 56 per cent in some parts of India where in some areas even electricity is supplied free for extracting underground water. In contrast, however, Latin America and Sub-Saharan Africa have 'large untapped water resources' for agriculture and with proper steps taken the supply of water for irrigation may be doubled by 2050 though still then some areas will remain water-stressed.

Apart from the effect of water availability on agricultural production another natural factor – aggravated by extremely irresponsible human action – known as climate change is going to hit hard on the productive system with serious impact not only on the production system making food globally insecure but also geographical shifting in the productive capacity of the crops.

First of all, global warming caused by higher degree of CO_2 emissions is likely to raise global temperature by 2°C-3°C in another 50 years or so making the situation worse with change in rainfall regime along with its associated elements of heavy rainfall or disastrous droughts. According to Stern (2009), 'average per capita CO_2 emissions across the world have increased around 15 per cent' during the last three decades. But it shows different trends in different regions during this period, for example average per capita emission of CO_2 has remained 'stable or slightly decreasing in high income countries' due to 'de-industrialisation of their economies with a population of 1.2 billion in 2007' while, on the contrary, average per capita emission of CO_2 has increased from 1 to 4 tonnes per year in the developing countries with a population about 5.2 billion, particularly in recent years but per capita emissions in least developed countries have been 'stable or very low' – around 0.2 tonnes per annum. World population increased during those three decades by 50 per cent resulting in an increase of CO_2 emissions worldwide by 'around two-thirds.'[22]

Stern (2009) stresses with greater emphasis the increase in global warming and its impact on physical environment intimately connected with human life and activities, including the sea level, precipitation rates, hurricane intensity and frequencies. According to Stern: 'One characteristic feature of climate change is higher variability on key dimensions. Record hot summers can be followed by cold spells, heavy monsoon rainfalls by intense dry periods. This in turn affects us humans – the fifth and final link in the chain.'[23]

A closer review of certain cases should dispel any doubt about the severity of effects of climate change. We can gather information and effect of temperature-rise from the study of ice cores recording evidence of CO_2 and temperature relationship. On the other hand, the effect of the sea-level rise are more easily available.A rise of sea-level by just *one metre* would directly affect 150 million people in Asia with its long exposed sea coast, exposing $1 trillion of economic assets. We need to be strongly reminded that the 'eventual sea-level rise from just 1°C will be several metres, not just one metre', and the rise of sea-level is an extremely lengthy process, once it starts there is no way to control it.

More severe dislocations will take place if the temperature rises by 2-3°C, such sea-level rises will then be accompanied by:

> a greater frequency of intense storms and hurricanes, the melting of glaciers accompanied by torrents and flooding in wet seasons, droughts in many parts of the world, and a high risk that the major rainforest will collapse. In the last two or three decades we have already seen some of the first forerunners of what might happen, and we are 'only' around 0.8°C above 1850 now.[24]

In this connection, many more formidable instances are provided by Stern connected with climate change triggered by rising temperature, like extensive forest fires (the most recent one in Australia), water reservoirs drying up, such as Lake Chad in Africa having shrunk from 25,000 square kilometres in 1963 to 1,350 sq. km. today. On the other hand, glacial melting and glacial lakes swelling in the Himalayan regions is going to create havoc when already glacial retreat in Bhutan has reached 30-40 metres per year and the temperature in the High Himalayas has already risen by about 1°C since mid-1970s causing glacial melting and resulting flooding along with landslides in the medium term turning into water shortages in the long term. These effects are going to have catastrophic effect on the regions which have so long been fed by the Asian river systems such as the Yangtse and Yellow Rivers in China, the Ganga, Brahmaputra and Jamuna in India and Bangladesh and the Indus in Pakistan. The loss of global meltwater would 'reduce July-September flow by two-thirds, causing severe water shortages for 500 million people and 37 per cent of India's irrigated land.'[25]

As reported in The World Bank Report (World Development Report 2008) effects of climate change are already quite visible in many regions, such as the Sahel becoming wetter while some other regions, like the Mediterranean and Southern Africa along with parts of Southern Asia are becoming drier. These changes are going to be very sharply reflected, especially in the agricultural activities of the affected regions, such as:

> With moderate warming, crop yields are expected to increase in temperate areas and decline in the tropics. Crop-climate models predict an increase in global crop production in slight to medium warming scenarios of less than 3°C. But the combined effects of higher average temperatures, greater variability of temperature and precipitation, more infrequent and intense droughts and floods, and reduced availability of water for irrigation can be devastating for agriculture in many tropical regions. One-third of the population at risk of hunger is in Africa, one-quarter in Western Asia and about one-sixth in Latin America.[26]

The comment made by The World Bank Report 2008 quoted above is very specific about the deteriorating situation of agricultural production in much of the tropical regions and felt most severely in Sub-Saharan Africa which may have to depend largely on imported food. However, this partly pessimistic view may change in future looking at the possibilities of Africa taking up this challenge and it is worth noting what Chairman Bingu was Mutharika of Malawi said at the assembly of African Union in February 2010:

> One challenge we all face is poverty, hunger and malnutrition of large populations. Therefore, achieving food security at the African level should be able to address these problems. I would therefore request the AU Assembly to share the dream that five years from now no child in Africa should die of hunger and malnutrition. No child should go to bed hungry. I realise that this is an ambitious dream but one that can be realised. We all know that Africa is endowed with vast fertile soils, favourable climates, vast water basins and perennial rivers that could be utilised for irrigation farming and lead to Green Revolution and mitigate the adverse effects of climate change. We can therefore grow enough food to feed everyone in Africa. I am, therefore, proposing that our agenda for Africa should focus on agriculture and food security. I propose that our slogan should be, 'Feeding Africa through New Technologies! Let us act now.'[27]

With reluctance, however, one has to admit that this has not yet been attained though some advances have already been made in a patchy manner, making agricultural progress quite admirable.

As a matter of fact, Africa is still essentially an agricultural continent when Sub-Saharan Africa's agricultural contribution is '34 per cent of GDP and 64 per cent of employment'. About 20 per cent of Africa's exports comprise agricultural products while application of agricultural research and extension services might make a 35 per cent return and another 15-20 per cent return through irrigation in Sub-Saharan Africa. It is not difficult to understand that agricultural development in Africa has not received similar impetus as in our countries, for example only 4 per cent of Africa's crops are irrigated making the continent highly vulnerable to weather shocks against 39 per cent in South Asia. Fertilizer use in Africa is less than 10 per cent of global average of 100 kilograms per hectare. Transport is virtually absent in many areas of Africa and 40 per cent of the rural population live in arid or semi-arid conditions. Similarly, 'about' 50 million people in Sub-Saharan Africa and 200 million in North Africa and Middle East live under acute water-stress. 'Cropland per agricultural population has been decreasing for decades' while 75 per cent of agricultural land suffers from 'excessive extraction of soil nutrient'.

Though agriculture in Africa accounts for 64 per cent of the labour force, over 34 per cent of GDP and over 20 per cent of businesses in most countries are in agricultural products, lack of investment has dwarfed its further progress. Further aggravating situations stalling agricultural progress have been racial conflicts, with more of them in the world happening in Sub-Saharan Africa though having abated in recent years. HIV/AIDS epidemic has acted as another reason stalling overall progress besides making for great disruptions.

The major reason keeping agricultural possibilities from coming into fruition is a lack of yield-practices like application of fertilizers, better seeds etc., are leading to Sub-Saharan Africa ranking 'lowest in the world' in these respects and as a result land productivity in Africa is 'estimated at 42 and 50 per cent of that in Asia and Latin America respectively'. Arable land in Africa amounts to '733 million hectares (27.4 per cent of World's total) compared with 628 million hectares in Asia and 570 million hectares for Latin America'. But sadly, only 3.8 per cent of African surface and ground water is used, covering 7 per cent of cropland (3.6 per cent for Sub-Saharan Africa).

Nevertheless, African agriculture has a huge potential with the application of advanced techniques and removal of other barriers which, properly handled, are likely to raise agricultural output from present US$ 280 billion per year to US$ 880 billion by 2030.

One of the most striking features of African agriculture is that a high proportion of the farmers are women, providing '70-80 per cent of the labour for food crops grown in Africa'. 'Female farmers make-up 48 per cent of the African labour force' but there is no gender parity when offering higher training as it is always dominated by men.[28]

VI. How to Tackle Climate Change

Controlling emissions of CO_2 is an extremely tricky problem especially because of lack of Governments' concern in many cases where they are more in the grip of giant corporates, or popularly called 'lobbies' in collusion with politicians, while in other cases there is a lack of knowledge about the right steps to be taken to mitigate climate change.

However, taking all these uncertainties into consideration we should at least have an idea how and to what extent one can go about it. According to Stern (2009), energy-related sectors are responsible for 63 per cent of global emissions while 37 per cent come from 'land-use change and deforestation' of which the latter and peat fires together are responsible for greater emissions than caused by all types of transport – road, sea, and air – taken together for two countries, Brazil and Indonesia. In fact, stopping deforestation will act as a great positive step forward in cutting emissions effectively.

Another important source of emissions is energy with electricity along with transport together contributing about 40 per cent of the world's total greenhouse gases in spite of the fact that electricity can be generated from a large group of sources with very low carbon emissions. They are – hydroelectricity, wind, solar, biomass, nuclear, geothermal, wave, tidal and, carbon capture and storage (CCS), mentioned in the previous Chapter. A number of countries have taken up strong initiatives in replacing fossil-fuels by some of these almost non-emission sources for supply of energy or power (electricity). In Iceland, for example, the major source of power is geothermal; Brazil and Scandinavian countries depend largely on hydro; in France electricity is 75 per cent nuclear, while Germany as the leading country in the world in using wind power varying between 30-50 per cent in some parts of the country, generating 23,000 MW in 2007. German manufacturers have 37 per cent share in the global turbine and component market, earning around 6 billion euros in exports in 2007. In contrast, however, China, India, Poland, and some others still derive 70-80 per cent of their electricity from using coal.

At the same time, the benefit to be derived from changing road transport, making it more public (bicycle) as discussed in previous Chapter is going to cut down emissions very effectively and has to start discouraging air transport so that emission of greenhouse gases going up from current 3-4 per cent of the total to up to 15 per cent by 2050 although Boeing estimates that 'biofuels could reduce flight-related gas emissions by 60-80 per cent.'[29*]

VII. Agricultural Productivity Gains Through Science and Technology

Now turning to how agricultural productivity gains have to be made, the emphasis as mentioned in The World Bank Report 2008 (Agriculture for Development), has to be laid on sustainable technology in the face of scarcity of land and water, making it workable even in regions like Sub-Saharan Africa not only for increasing productivity but at the same time making the system more stable and resilient. Such an approach will not only raise yields but at the same time save energy and water-use, reduce risks of crop-loss as well as 'improve product-quality, protect the environment and make female participation more gainful'.

Such an approach based on science and technology will act positively in reducing 'yield-losses, increase famers' profits, greatly reduce use of pesticides for millions of small holders but also help in finding a way to make commercial agriculture practised by private multinationals useful for serving 'the needs of the poor'.

It is, however, altogether a different approach when use of science and technology is made specifically for agriculture – 'a biological process'. Research conducted since the 1960s succeeded in developing improved varieties of animal breeds adapted to subtropical and tropical conditions making high returns with pro-poor impacts, promising even better varieties in near future.

Since the 1960s, scientific plant breeding has developed 'improved varieties of seeds' suited to small holder and sub-tropical and tropical areas' constituting 'semi dwarf varieties of rice and wheat and improved varieties of maize' and have contributed to yield growth with greater impact than the Green Revolution, accounting for 'as much as 50 per cent yield growth' without which world food prices would have been '18-21 per cent higher in 2000' making '13-15 million children malnourished'. The genetic improvement suitable for both irrigated and rain-fed crops has continued to provide yield gains and already these improved varieties of seeds are planted in '80 per cent of cereal area in India' half of which is irrigated, and globally the area planted with them has 'more than doubled since 1981' increasing yield by 1 per cent, largely in rain-fed areas. Subsequently, even Sub-Saharan Africa as well is receiving an impact from these biological breeding techniques at least with some major crops like maize, improved varieties of disease-resistant strains of cassava, rice and beans combining the high-yielding potential of Asian rice with the resistance of African rice to 'weeds, pests and diseases and water-stress increasing yield now cultivated on 2,00,000 hectares in Africa and new varieties of beans with multiple stress-resistance grown widely in eastern, central and southern Africa by about '10 million farmers, mostly women.'[30]

In short, the major emphasis laid by R&D is on 'improving the management of crop, livestock, and the natural resource system.' Much of this work is directed to soil and water management along with agroecological approaches stressing the reduction of agricultural chemicals. The steps to be taken in this direction primarily consist of 'conservation, or, zero tillage, maintaining crop residues as ground cover, including many other advantages like saving labour and energy, increasing drought tolerance, and reducing greenhouse gas emissions. In Latin America, Brazil and Argentina together use zero tillage on more than 40 million hectares or 43 per cent of the arable land and which has even spread to small farmers in Southern Brazil. It has also been adopted by small farm holders in Ghana, and further is going to be rapidly adopted by small farmers on 0.8 million ha in the Indo-Gangetic Plain including various parts of the technology such as 'rotation, seeds, chemicals, and machinery'. Another very important step towards 'biologically fixing nitrogen to the soil is the introduction of legumes while in South Africa fast-growing 'fertilizer trees such as Gliricidia, Sesbania, and Tephrosia' not only improve soil-quality but also check soil erosion. At the same time, improved methods of pest management with reduction in the use of dangerous pesticides along with the use of legumes replacing chemical fertilizers coupled with 'rotational fallow' have 'more than doubled' maize yield in Southern Africa. According to the 'Consultative Group on International Agricultural Research Science Council', the practice of rotational fallows and permanent intercropping have benefited 12 million small holder maize farmers in Eastern and Southern Africa protecting soil fertility, keeping it 'productive and healthy.'[31]

Hence, undeniably, agricultural backwardness, its failure to fight most of natural elements can be tackled with application of researches on technological approaches which are closely linked with greater emphasis on larger investments as guiding principles for future approaches.

However, it is very discouraging to note the great reluctance of the governments or people in power to invest in R&D contributing directly to agricultural production and thus stem hunger and keep the society healthy. According to The World Bank Report 2008, as a group the developing countries 'invested 0.56 per cent of their agricultural gross domestic product (GDP) in agricultural R&D in 2000 in contrast to 5.16 per cent invested by developed countries. A few developing countries, however –'China, India and to a less extent Brazil – have rapidly increased their spending on agricultural R&D over the past two decades'. But Sub-Saharan Africa's initiative in this regard is far from praiseworthy particularly because she lacks the opportunity of spill-over from other developing countries' attainments since her major food crops – cassava, yams, millets, plantain and teff are entirely different while investment connected with R&D in other developing countries does not match with Sub-Saharan Africa's requirements. Besides, so far the spending per scientist in Sub-Saharan Africa is low in spite of having '50 per cent more scientists than India and about a third more than the United States' (Table 4.2).

Understandably, the small holder farming system, the most prevailing system providing food for own and others' consumption, suffers from lack of access and inputs, credit, and incapability to bear risks particularly connected with climate change. In order to tackle these challenges, the small holder farmers are encouraged to stress demand-led agricultural extension taking help from agricultural research and exchanging information on seeds on the basis of farmer-to-farmer networks. In other words, such development replicates La Campesina approach discussed earlier as fruition of Food Sovereignty movement.

Table 4.2: Comparison of Agricultural Research Systems in Sub-Saharan Africa, India and the USA around 2000

	Sub-Saharan Africa	India	United States
Arable and Permanent Crop Area (hectares/millions)	147	160	175
Number of Public Agricultural Research Agencies	390	120	51
Number of Full Time Equivalent Scientists	12,224	8,100	9,388
Percentage of Scientists with Ph.D.	25	63	100
Annual Public Spending on Agricultural R&D (1999 int'l $ millions)	1,085	1,860	3,465
Spending per Scientist (1999 int'l $ thousands)	89	230	370

Source: *The World Bank: World Development Report 2008*, p. 168.

Needless to say, that all these exercises are finally aimed at making food available for everybody, especially for those, the small holder farmers in the developing countries, who grow it in anticipation of meeting their own hunger and providing it for others who need it, in other words, having the scope for selling it in the market at a reasonable price to be used for meeting other essential needs of the family. In short, the entire exercise beginning from production to sale, acts as a process giving him/her freedom from hunger.**

It is, however, yet far from clear if the world resources are capable of feeding all the people particularly whether these small holder farmers produce enough for their own consumption, as well as for the consumption of a global population still growing in number.

This question arises from the doubt in the minds of many experts in the field and to dispel which we may turn our attention to the monumental work of the most recent Nobel Laureates (jointly) Abhijit V. Banerjee and Esther Duflo who are very categorical about 'poverty trap' which, in their opinion, 'might not' exist.

> At least in terms of food availability, today we live in a world that is capable of feeding every person that lives on the planet ... starvation exists in today's world, but only as a result of the way the food gets shared among us. There is no absolute

> scarcity … it seems that most people, even most very poor people, earn enough money to be able to afford an adequate diet, simply because calories tend to be quite cheap, except in extreme situations.[32]

According to them 'most people earning less than 99 cents, that is $1 a day do not behave 'as if they are starving' otherwise they would have spent all of it buying food. However, the data collected by the authors from 18 countries show that 'from 36 to 79 per cent' of the money is spent by extremely poor rural people on food. In specific cases from their findings in Udaipur, India, it was found that 30 per cent more could be spent on food if the poor cut expenditures on 'alcohol, tobacco and festivals'. At the same time instead of spending money that they spend on nutritious food much is spent on buying 'better-tasting, more expensive' food. As observed by the authors spending large amounts on 'weddings, dowries and christening' makes a very important part of life, even among extremely rural poor to 'avoid a loss of face', including large spending on funerals of the elderly in South Africa so that the King of Swaziland 'banned lavish funerals' by law while, in India, poor people spend '14 per cent of their budget on festivals'. It is almost a universal feature that buying a television is 'more important than food' among the poor to make 'life less boring'. What is, however, more alarming according to Banerjee and Duflo (2013) is malnourishment of children among the poor everywhere. As reported by the National Health Survey (NFHS-3), in India almost 'half of the children under five are stunted' which means that they are far below the norm in weight. One-fourth of them are severely stunted, which is, in fact, double the 'wasting rates' of Sub-Saharan Africa, the poorest part of the world.[33]

> Childhood malnutrition directly affects the ability of adults to function successfully in the world … proper nutrition in childhood has far-reaching implications … undernourished children are more likely to become short adults, to have lower educational achievement and to give birth to smaller infants. Undernourishment is also associated with lower economic status in adulthood.[34]

As a matter of fact, lacking the knowledge of how to feed children mostly among the poor class happens to be a very important factor in leading to such a situation. For example, according to a report in the *British Medical Journal 1995*, mothers taking iodine capsules during pregnancy led to a '7.5 per cent increase in the total educational achievement of their children'. It is suggested that to overcome this problem of malnutrition among children, the food-producers should put greater stress on producing 'new strains of nutritious and tasty crops that can be grown in a wider range of environments' treating it as a priority for food technology with similar stress on raising productivity.[35]

At the same time society should take steps towards taking care of women in pregnancy so that BMI (Body Mass Index) improves in the long run in the absence of which the women remain short such as the Nepalese, Indian and Bangladeshi women, the shortest ones in the world, and as such are very likely to bear small infants.

This discouraging situation of the poor, especially in the developing world, leads to the ultimate question whether the poor in rural areas hold any prospect of facing the great challenge of feeding the growing population in future. Jules Pretty (2010), rather, feels quite optimistic about it provided future steps are taken in the right direction. For instance, though there is enough staple food produced in the world 37 per cent of it is fed to animals in the developing countries and 73 per cent in industrialised countries, along with a substantial part of it wasted by the rich during consumption. In order to overcome these wrong policies, a new approach to the productive process is advocated in which 'locally available and/ or cheap technologies combined with making the best natural, social and human resources' using 'best available technologies along with best ecological management' will give rise to a new approach which may be adopted as *'sustainable intensification'* causing minimum harm to the environment. The major traits of its practice involve using nitrogen-fixing legume replacing fertilizers, and in place of using pesticide putting emphasis on natural enemies and reduced tillage (or no tillage) besides 'making better use of land, water and biodiversity and technologies.'[36]

Jules Pretty (2010) in this connection recommends a number of approaches to 'resource-conserving technologies'.*** and ****). However, implementation of all these practices makes it essential for farmers to be able to acquire the management skill. It is now highly encouraging to know that 'some of the most significant progress' has already been achieved in the developing countries towards the successful implementation of sustainable agroecosystems during the past decades results of which are revealed in the analysis of data of 286 such projects spread over 57 countries, engaging some '12.6 million farmers on 37 million hectares in transition towards agricultural sustainability' though representing 'just over 3 per cent' of the total cultivated area (1.136 million hectares) in developing countries. Farmers engaged in wetland rice-based systems, mainly in Asia represented the largest number and Southern Latin America contained the largest area engaged in mixed system. (See Table 2.2. included in Chapter 2).

Juies Pretty (2010), thus concluded that though we are not feeding all the people in the world, it may be possible to feed the estimated 9 billion people by the mid-century provided full success is attained in the adoption of sustainable intensification of agriculture, which is likely to be highly promising, especially for the poorest, The International Assessment of Agricultural Knowledge, Science and Technology for Development concluded that: 'a focus on small-scale sustainable agriculture, locally adapted seed and ecological farming better address the complexities of climate change, hunger, poverty and productive demands on agriculture in the developing world.'[37]

Let us hope that the role of solving global problems of food may be tackled by the small holder rural farmers who are not likely to change their agricultural profession developed through generations with an intense attachment with the soil (though their children given a common effect of education may seek livelihood in urban areas).

NOTES

***Stern, Nicholas.** *A Blueprint for a Safer Planet: How to Manage Climate Change and Create a new Era of Progress and Prosperity*. The Bodley Head, London; 2009. p. 53.

'Consider wind energy, for example. Ever since antiquity people have sought to use the power of wind to lighten the burden upon themselves. Around AD 600, the Persians designed simple windmills to grind grain. In the twelfth century, the famous Dutch windmills were used to drain water from fields. The windmills of pre-industrial Europe were estimated to provide 1,500 megawatts of power (a level not reached until the 1980s). In 2007, the top five countries in terms of installed capacity are Germany (22.3 GW), the U.S. (16.8 GW), Spain (15.1 GW), India (7.8 GW) and China (5.9 GW). In terms of economic values, the global wind market in 2007 was worth about 25 billion euro ($37 billion) in new generating equipment. The Spanish government has a target to install a total 20 GW of wind power by 2010 – wind farms already account for 10 per cent of electricity generated in Spain in 2007. The US Department of Energy has set a target of generating 20 per cent of its energy from wind power by 2030, totalling 300 GW of electricity generation. This means having capacity from 2000 installations per year in 2006 to almost 7000 per year in

2017. The department asserts that integrating 20 per cent wind energy into the grid can be done reliably for less than 0.5 per cent per Kwh. Wind accounted for 35 per cent of total installed power capacity in the US in 2007'.

**Sen, Amartya. *Food and Freedom. World Development* 17. 1989. p. 91.

'If a person is not free from hunger and lacks the means and practical opportunities to feed himself or herself adequately, then that person's positive freedom must be seen as having been thoroughly compromised'.

***Pretty, Jules. "Can Ecological Agriculture Feed Nine Billion People?" in Magdoff F., Tokar B. (Ed). *Agriculture and Food Crisis: Conflict, Resistance, and Renewal.* Monthly Review Press, New York; 2010. pp. 289-290.

'There are many types of resource conserving technologies and practices that can be used to improve the supplies and use of natural factors in and around agroecosystems. Some include: (i) *Integrated Pest Management*, which uses prevention through developing ecosystem resilience and diversity for pest, diseases and weed control, and only uses pesticides when other options are ineffective. (ii) *Integrated Nutrient Management*, which seeks both to balance the need to fix nitrogen within farm systems with the need to import inorganic and organic sources of nutrients, and to reduce nutrient losses through control of runoff and erosion. (iii) *Conservation Tillage*, which reduces the amount of tillage, sometimes to zero, so that soil can be conserved through reduced erosion, and available moisture is used more efficiently. (iv) *Cover Crops*, which grow in the off-season or along with main crops, help protect soil from erosion, manage nutrients and pests, maintain healthy soil, enhance water infiltration and storage in soil. (v) *Agroforestry*, which incorporates multifunctional trees into agricultural systems, and collectively manages nearby forest systems. (vi) *Aquaculture*, which incorporates fish, shrimp, and other aquatic resources into farm systems, such as irrigated rice fields and fishponds, and so leads to increases in protein production. (vii) *Water Harvesting* in dry land areas, which can mean that formerly, abandoned, and degraded lands can be cultivated, and additional crops can be grown on small patches of irrigated land owing to better rainfall retention. (viii) *Livestock Reintegration* into farming systems, such as raising the dairy cattle, pig, and poultry, including using both grazing and zero-grazing cut-and-carry systems. Mixed crop-livestock systems provide many synergies that enhance production and allow for better nutrient cycling on farms'.

****Pretty, Jules. "Can Ecological Agriculture Feed Nine Billion People?" in Magdoff F., Tokar B. (Ed). *Agriculture and Food Crisis: Conflict, Resistance, and Renewal.* Monthly Review Press, New York; 2010. p. 290.

'The individual practices are focused on growing healthy plants with good defence capabilities, enhancing beneficial organisms and stressing pests, while maintaining or enhancing environmental quality.... For example, hedgerows and alley crops encourage predators of pests and act as windbreaks, thus reducing soil erosion by wind. Legumes introduced into rotations of fix nitrogen, and also act as break crop to prevent carryover of insects, pests and diseases. Grass contour strips slow surface water runoff, encourage percolation of groundwater and can be a source of fodder for livestock. Cover crops prevent soil erosion and leaching during critical periods and can also be plowed in as a green manure. The incorporation of green manures not only provides a readily available source of nutrients for the growing crops but also increases soil organic matter, and hence water retentive capacity, further reducing susceptibility to erosion'.

1. *United Nations World Resources 2000-2001*. People and Ecosystems: The Fraying Web of Life. United Nations Development Programme. United Nations Environment Programme. World Bank, World Resource Institute. p. 53.
2. *Ibid.*, pp. 56-57. FAO (2000) refers to Statistical Databases of the Food and Agriculture Organisation of the United Nations.
3. *Ibid.*, pp. 56-60.
4. *Ibid.*, p. 68.
5. *Ibid.*, pp. 69-73.
6. *Ibid.*, p. 75.
7. *Ibid.*, p. 99.
8. *Ibid.*, pp. 119-123.
9. The World Bank: Development Report 2003. Sustainable Development in a Dynamic World, Transforming Institutions, Growth, and Quality of Life. p. 83.
10. Wittman H, Aurelie A.D., Wiber, N. "The Origin and Potential of Food Sovereignty", in *Food Sovereignty: Reconnecting Food, Nature and Community,* ed. H. Wittman, A.D. Aurelie and N. Wiber. U.K.: Pambazaka Press, 2010, p. 3.
11. *The World Bank: World Development Report 2003*. Sustainable Development in a Dynamic World, Transforming Institutions, Growth, and Quality of Life. pp. 83-87.
12. *Ibid.*, pp. 86-90.
13. *Ibid.*, pp. 105-106.
14. *The World Bank: World Development Report 2008*. Agriculture for Development. pp. 50-53.
15. *Ibid.*, pp. 58-60.
16. *Ibid.*, p. 60.
17. *Ibid.*, pp. 61-62.
18. Dyson, Tim. *Population and Food: Global Trends and Future Prospects*. London: Routledge, 1996, pp. 115-117.
19. Ehrlick, P.R., Ehrlick, A.H., Daly, G.C. *Population and Development Review* 19.1 (1993), pp. 2, 7.
20. *Ibid.*, pp. 6-7.
21. International Food Policy Research Institute (IFPRI). 2002 Summary. p. 2.
22. Stern, Nicholas. *A Blueprint for a Safer Planet: How to Manage Climate Change and Create a New Era of Progress and Prosperity*. London: The Bodley Head, 2009, p. 20.

23. *Ibid.,* p. 27.
24. *Ibid.,* p. 29.
25. *Ibid.*
26. *Agriculture for Development.* The World Bank: World Development Report 2008, p. 65.
27. Juma, Calestous. *The New Harvest: Agricultural Innovation in Africa.* Oxford: Oxford University Press, 2011, p. xiii.
28. *Ibid.,* pp. 7-8, 11-16, 116, 118.
29. Stern, Nicholas. *A Blueprint for a Safer Planet: How to Manage Climate Change and Create a New Era of Progress and Prosperity.* London: The Bodley Head, 2009, pp. 42-43, 45-46 and 115.
30. *The World Bank: World Development Report 2008.* Agriculture for Development. pp. 159-160.
31. *Ibid.,* pp. 163-165. See Consulative Group on International Agricultural Research Science Council, 2006.
32. Banerjee, A.V. and Duflo, E. *Poor Economics: Rethinking Poverty & Ways to End It.* India: Penguin Books, 2013, pp. 36-37.
33. *Ibid.,* pp. 32-33, 43, 51-53.
34. *Ibid.,* p. 46.
35. *Ibid.,* p. 58.
36. Pretty, Jules. "Can Ecological Agriculture Feed Nine Billion People?" in *Agriculture and Food Crisis: Conflict, Resistance, and Renewal,* ed. F. Magdoff, B. Tokar. New York: Monthly Review Press, 2010, pp. 284-285 and 288.
37. *Ibid.,* p. 298.

5
The Inherent Strength of Rural Agriculture in Meeting Global Hunger for Food

I. A Dreadful Situation Facing Mankind

We may begin to analyse this challenging situation referring first to George Monbiot writing for the *Guardian*, UK, in the Journal Column on Monday 11 December, 2017. He writes about being terribly upset to realise the growing global situation precipitating mass starvation by the middle of this century when the global population by then will increase by two or three billion. The reason behind the development of such a terrible situation is soil loss through degradation during the next 60 years when yields are already declining on 20 per cent of 'world's croplands'. Added to this is the loss of water as in places like the North China Plain, the central United States, California, and north-western India where the levels of groundwater used for irrigation have already reached a critical level, such as in Upper Ganges basin where water is already drawn '50 times its recharge rate' while the farmers in South Asia expect to use '80 and 200 per cent more water by 2050' though nobody knows where it will come from (Monbiot). Growing temperatures become the next restraint when all remaining the same, each degree Celsius it increase slowers the yield of rice by '3 per cent, wheat by 6 per cent, and maize by 7 per cent' and when an increase of temperature by 4°C would reduce maize production 'by 84 to 100 per cent'.

In this very gloomy situation, however, the only brighter side is offered by small, rural farmers putting in more labour, growing a wider range of crops, working land more carefully, though owning less than five hectares each (when they own any land) and working only 30 per cent of the cropland grow 70 per cent of the food, when fertile land twice the size of England has already been grabbed and converted by Corporate Houses into large farms growing crops for export, as mentioned in the previous chapter, instead of feeding the poor.

On the other hand, the disaster in the food-situation has multiplied by the extremely worsening condition of the seas becoming the repository of plastic rather than fish, the catch declining '1 per cent a year' depriving living to about 3 million people largely dependent on fish and shell-fish protein. Contrastingly, however, global meat production has increased four times in the last 50 years though its consumption per head worldwide is half of that in the UK, and only a third that in the USA. In consequence as Monbiot (2017) reports, 'already 36 per cent of the calories available from grain and pulses and 53 per cent of the protein' are used to feed animals 'to provide animal protein for human consumption'. In this way, more and more land is turned into grazing grounds for increasing meat production and according to the US calculation the meat consumption (obviously by the rich) made at the current rate will rise by '70 per cent' by 2030, doubling its demand for crops by 2050 than what it was in 2005 when the land required to meet this demand does not exist (Monbiot). The only visible solution to this problem, as suggested elsewhere before, seems to be a shift from an animal-to-plant-based diet, stopping meat production and stopping the use of food-crops for the production of biofuels which 'could provide food for another 4 billion people doubling the protein for every human consumption.'[1]

II. Food Consumption, the Changing Pattern, and Food Availability

Looking from another angle we may, perhaps, become partly relieved that since global food production grew 'faster than population and per capita consumption increased' (Alexandratos and Bruinsma, 2012) – an increase (expected) of 2.25 billion between 2010 and 2050 in comparison to 2.5 billion in 1950 to 3.7 billion in 1970 – and thus decelerating the impact of world agriculture by 'lowering its growth'. According to them, there is sufficient food to satisfy the needs of everyone 'to be well fed', though this did not happen, leaving about 2.3 billion people to survive on 2500 kcal per day instead of 2770 kcal available in 2005/2007, along with some 0.5 billion with less than 2000 kcal though at the other extreme some 1.9 billion were consuming more than 3000 kcal. The reasons of leaving a large section of people, consisting of poorer section of mostly rural population half-fed is partly the result of using food-crops, as mentioned earlier, for the production of biofuels, as for instance, in: 'the case of maize used for ethanol in the US, of sugarcane in Brazil, of vegetable oils and cereals in the EU to produce biodiesel and ethanol.'[2]

As a matter of fact, 'population growth, increase in per capita consumption of food, and consumption of more livestock product' needed an increase in the demands for agricultural products to grow at 1.1 per cent per annum from '2005/2007 – 2050 in comparison to 2.2 per annum during the past four decades.

The most important fact about the nature of food consumption, which, in effect guides the policy of agricultural development of small holder farmers is the composition of food consumed, such as meat and milk comprising substantial part of the meals taken in the developed countries and likely to make a greater proportion in future, while it is not the same with the developing countries in spite of the growing popularity of meat and milk, though the same largely hinging on rising income as well as general habits of the people, is very much applicable to the difference in consumption pattern in China with its growing love for meat-based diet contrasted with that in India where vegetables and pulses (may be including, fish and some poultry) predominate in some of its parts.

In other words, any visible change in the consumption habits of the largest section of the people living in the Global South is not imminent. On the other hand, the commitment of The 1996 Food Summit to reduce the number of undernourished – obviously in the Global South – by '50 per cent by 2015' was not fulfilled which is not expected to be accomplished 'before the second half of the 2040s'. The somewhat encouraging feature acting as a partly mitigating feature against improving consumption situation is that cereal production in the world is going to make a substantial increase by 940 million tonnes over that of 1,225 million tonnes between 1961/63 and 2005/2007, and expected to reach a total of 3 billion tonnes by 2050.[3]

A situation like this is going to exert visible changes in trading, such as the developing countries have so far been traditionally net importers of cereals like wheat and coarse grains while exporters of rice. Among the great importers were Mexico, Saudi Arabia, the Republic of Korea, Egypt, Algeria, and Taiwan while, meanwhile, export from some of the developing countries – like China and India exporting rice in ever larger quantities to South America and the export of coarse grains from China and even wheat from India are noteworthy.

Significantly, however, the constraints for maintaining agricultural production are land and water particularly in the face of declining yields but the former – that is availability of more land and water – appear to be the major hindrance in the way of agricultural expansion, as discussed in the previous chapter. The current situation reminds us about the prediction made by the great English Economist of his time, Thomas Malthus, who could oversee the consequences of population growth exceeding the food supply ultimately leading to insecurity explained clearly in his book 'Essays of Population' written in 1798, giving opposite examples of geometrical progression of population against an arithmetical progression of food,[4]

In this respect, Alexandratos and Bruinsma (2012) looking for information on present status of factors encouraging agricultural expansion in future, in other words strengthening food security on the basis of information gathered from sources like IIASA, FAO and Global Agro-Ecological Zones Study (GAEZ) found out that significant amount of land 'with rain-fed production potential is available, amounting to 7.2 billion ha of which 1.6 billion ha is already in use including irrigated ones'. According to GAEZ, another 75 million ha also used are unsustainable for production. Of the remaining 5.7 billion ha, 2.8 billion ha are under forest cover or already are in use for settlements, infrastructure etc., leaving only about 1.4 billion ha of 'prime land' and 'good land' that could be brought into cultivation depending on development of proper infrastructure at considerable costs besides fighting diseases etc. (as in Sub Saharan Africa).

This undoubtedly, shows the limitations of more land available for agricultural expansion. This, apparently, gives no other option left for increasing agricultural production except through raising yield. Land needed to meet the required amount of food is '70 million ha' and the prospect of getting it means '132 million ha expansion' in countries like Sub-Saharan Africa and Latin America while at the same time along with decrease of land under crops in developed countries and with some even developing countries. It is, however, pointed out that there are thirteen countries, namely Madagascar, Mozambique, Canada, Angola, Kazakhstan, the Democratic Republic of Congo, China, the Sudan, Australia, Argentina, Russian Federation, the United States of America and Brazil, put in ascending order, providing 60 per cent of the 1.4 million ha of 'prime' and 'good' land which is not being used for any purpose and could be converted for agricultural expansion, though may be involving high costs.

The second major constraint for agricultural expansion is availability of water for irrigation. According to the report of Alexandratos and Bruinsma (2012), areas equipped for irrigation in the world stands at present at 300 million ha, more than double of that in early 1960s, but 80-90 per cent of it is under irrigation today concentrated mostly in the developing countries accounting for about 40 per cent of their harvested area, growing 60 per cent of their cereals, nearly half of the irrigated area concentrated in China and India besides holding the prospect of increasing one-third of the projected increase. On the other hand, renewable water resources (which is defined 'as the sum of annual precipitation and net incoming flows – consisting transfer through rivers from one to another – minus evapotranspiration and groundwater recharge) are extremely scarce, amounting to hardly 6.6 per cent globally, though in Near East/ North Africa and South Asia they accounted for 52 per cent and 40 per cent respectively in 2005/2007.

As we have discussed earlier, there is no real constraint to produce enough food to meet much greater requirements of a larger population by 2050 apart from the real constraint of their proper supply system which obviously includes storage, transport, distribution without any hindrance. This involves much more serious as well as honest role of the local government than any private agency to allow easy access especially of poor people to available food.

It may imply the amount of food production needed to meet requirements. Since population growth shows declining signs in many countries, such as in Japan, Europe, China and Brazil (as evidenced in, Tables 3.8 and 3.9, in Food ; The Growing Challenge for Human Survival, 2020, Bimalendu Bhattacharya) many countries of Latin America and the Caribbean countries, which provide definite instances of population decline between 2015 sand 2050.[5]

All these factors naturally strengthen the view that the requirement of the amount of food needed to satisfy the hunger for a larger population by 2050 may grow at a conceivably much lower rate than in the past, such as the total agricultural output will come down to 1.1 per cent per year during 2005/2007 to 2050 from 2.2 per cent per year during the preceding equal period.

However, in spite of lower rate of production, cereal production would increase by 940 million tonnes to '3 billion tonnes by 2050', total meat production would be 455 million tonnes increasing by 196 million tonnes by 2050 while oil crops would be 282 million tonnes increasing by 133 million tonnes by 2050. All this conjectural progress would happen provided necessary steps, such as proper investments and 'right policies' providing incentive for farmers are given.[6] Another very important development, as mentioned in the previous chapter, is the conversion of increasing extremely harmful production of food and oil crops for biodiesel and ethanol production which will be for human food-security.

III. Availability of Food Nutrition in the Perspective of Changing Global Population

World population growth peaked in 1960s at 2 per cent per year and then fell to 1.2 per cent per year in the decade ending in 2010 followed by, as predicted, deceleration henceforward to 0.4 per cent per year by 2050, bringing total population to 9.2 billion by 2100. In other words, despite the 'drastic fall in growth-rate', the annual addition to total population amounted to 79 million till 2010, is expected to be 50 million every year till mid-2030s and then is expected to decelerate to 27 million per year by 2050. As a matter of fact, it is in the developing countries that all increases are confined while actual decline will be happening in the developed countries. However, the rate of increase will be widely different, such as East Asia will start recording negative population growth by early 2040s while negative growth will start in Latin America in the early 2060s and in mid-2060s in South Asia, and in Near East/North Africa in the mid-2080s. Thus, all these regions will start losing, while in stark contrast Sub-Saharan Africa, the only region in the world to go on adding to 2100 some 11 million each year reaching a total of 2.2 billion by 2100.[7] The trend of world population growth obviously dominated by a declining trend from 2030 onwards, makes the goal of ending hunger more difficult to achieve as long as the overall situation in Sub-Saharan Africa does not improve.

In this connection, it is necessary to get an idea about the composition of food consumed by the people in general, and the most striking feature in this connection is the rapid change in composition with increasing consumption of 'livestock meat (popularly called red meat), milk, eggs and vegetable oils'. A clear idea may be obtained in this regard from the accompanying Table 5.1 which brings out difference in the commodity composition between different worlds: Developed and Developing. Finally, we will consider the necessary changes in the agricultural production.

(1) The first notable difference in the consumption pattern of food between the Developing and Developed world happens in 'cereals for all uses' which is around three times higher per person all through the period 1969/71-2050 in the Developed countries as compared with that of Developing countries though the same does not happen where cereals (like rice, wheat, maize, millets etc.) are taken for consumption which, in other words mean that 'cereals for all uses' mean they are consumed not particularly for human consumption but as well as consumption for other purposes, such as animal feed producing more meat, ethanol and biodiesel for industrial purposes.

(2) Consumption of sugar as well is much higher in Developed than Developing Countries per person which is directly connected with growing obesity while people in the developing world are financially handicapped (in terms of lowpurchasing power) to consume more sugar.

(3) At the same time vegetable oils have a growing popularity in Developed Countries being preferred to the use of butter for cooking.

(4) Though in terms of kcal/per person/day, the consumption of meat has increased over the years in the developing world, as mentioned previously, the developed countries per capita/day use of it is more than double.

(5) However, the most spectacular difference between the two worlds in the consumption (or preference) for meat as a part of food is quite astounding, higher by over three times in the developed world till 2005/2007 after which the gap becomes less as meat is also hugely preferred as a part of daily food even in the developing countries (particularly in China but very little in India) though even by 2050 per head meat consumption in the Developed World will remain more than double that in the developing world.

(6) Similarly, milk and dairy products (excluding butter) are consumed nearly six times higher in the Developed World compared with that in the Developing World. In spite of its increasing share in daily food consumption of the Developing World, it remains about three times higher (kcal/person/day) in the Developed World.

(7) Projections from this data show, food consumption, taking all major types of food, remains consistently lower measured in Kcal/person/day increasing from 2056 to 2654 in 1969/1971-2050 in the Developing World while it is 3138-3490 in the Developed World during the same period with systematic increase all through the period.

On the whole, the major point largely overshadowing other changes in the consumption of food is:

> the real force behind the structural change has been growth in consumption of livestock products in countries like China (including Hong Kong SAR), the Republic of Korea, Malaysia, Chile, Brazil and Saudi Arabia. Indeed … the increase of meat consumption of developing countries from 14 to 28 kg from 1979/1981 to 2005/2007 was decisively influenced by the rapid growth in China and Brazil. Excluding them from the totals, the average of other developing countries

grow much less over the same period, from 12 kg to only 17 kg' (Table 5.1).[8]

It cannot be denied that agriculture in its present form faces enormous challenges both from providing food security as well as one of the most serious environmental risks prompting climate change. Such a dual role, dramatically opposite to each other is far from easy to solve to enable agriculture to feed the growing global population as well as maintain environmental sustainability too.

According to FAO, 'croplands cover 1.53 billion (about 12 per cent of ice-free land) hectares while pastures cover another 3.38 billion hectares (26 per cent of Earth's ice-free land), together representing the 'largest use of land' on the earth's surface, expanding further by 154 million ha between 1985 and 2005 increasing crop production by 28 per cent during the same period. Interestingly, increase in cropland during this period gave a larger yield – 'nearly three times the change in cropland area, resulting from increased multiple cropping, fewer crop failures, and less land left fallow', though taking increase in harvested land, crop yields increased by 20 per cent between 1985 and 2005.[9]*

According to J.A. Foley *et al.* 2011, re. non-food uses of crops, almost 35 per cent of crops are used indirectly for human use, such as using them as animal-feed, (for increasing the production of meat) or 'seed, bioenergy and other industrial products which reduces the amount of food directly available for human consumption to 62 per cent while 3 per cent is for bioenergy. At the same time, there is great disparity in the use of land for growing crops directly for human consumption such as, according to these authors, North America, and Europe use only 40 per cent of their cropland for producing crops for direct human consumption against 80 per cent in Asia and Africa (with the extremes of 25 per cent in the USA and over 90 per cent in South Asia). Further, quite frustratingly 75 per cent of the World's agricultural land is for raising animals, resulting, in fact, in a 'net drain on world's potential food supply'.

Table 5.1: Changes in Commodity Composition of Food by Major Country Groups

Kg/ Person/ Year	**1969/ 1971**	**1979/ 1981**	**1989/ 1991**	**2005/ 2007**	**2030**	**2050**	**Compa-rison 1991- 2001**	
							New	**Old**
				World				
Cereals/ food	144	153	161	158	160	160	158	165

contd...

Kg/ Person/ Year	1969/ 1971	1979/ 1981	1989/ 1991	2005/ 2007	2030	2050	Compa-rison 1991-2001	
							New	Old
Cereal/all uses	304	325	321	314	329	330	309	309
Roots and Tubers	84	74	66	68	77	77	69	69
Sugar and Sugar Crops (raw sugar eq.)	22	23	22	22	24	25	23	24
Pulses, dry	7.6	6.5	6.2	6.1	6.5	7.0	6.0	5.9
Vegetable Oils, Oilseeds and Products (oil eq.)	7	8	10	12	14	16	11	12
Meat (Carcass Weight)	26	30	33	39	45	49	37	37
Milk and Dairy, excl. butter (fresh milk eq.)	76	77	77	83	92	99	78	78
Other Food (Kcal/ person/ day)	194	206	239	294	313	325	285	289
Total Food (Kcal/ person/ day)	**2374**	**2497**	**2633**	**2772**	**2960**	**3070**	**2719**	**2789**

contd...

Kg/ Person/ Year	1969/ 1971	1979/ 1981	1989/ 1991	2005/ 2007	2030	2050	Compa-rison 1991- 2001	
							New	Old
Developing Countries								
Cereals/ food	140	152	160	155	159	158	157	166
Cereal/all uses	193	219	229	242	254	262	239	238
Roots and Tubers	79	70	62	66	73	78	67	67
Sugar and Sugar Crops (raw sugar eq.)	15	17	18	19	22	24	19	21
Pulses, dry	9.3	7.8	7.3	7.0	7.4	7.7	6.8	6.7
Vegetable Oils, Oilseeds and Products (oil eq.)	4.9	6.4	8.4	10.1	13.1	15.4	9.4	10.4
Meat (Carcass Weight)	11	14	18	28	36	42	26	27
Milk and Dairy, excl. butter (fresh milk eq.)	29	34	38	52	66	76	45	45
Other Food (Kcal/ person/ day)	115	130	177	253	279	293	242	242

contd...

Kg/ Person/ Year	1969/ 1971	1979/ 1981	1989/ 1991	2005/ 2007	2030	2050	Compa-rison 1991-2001	
							New	Old
Total Food (Kcal/ person/ day)	**2056**	**2236**	**2429**	**2619**	**2860**	**3000**	**2572**	**2654**
Developed Countries								
Cereals/ food	155	156	162	167	166	166	163	163
Cereal/all uses	571	620	618	591	682	695	564	565
Roots and Tubers	96	84	78	77	73	72	78	78
Sugar and Sugar Crops (raw sugar eq.)	41	40	36	34	33	33	33.6	34
Pulses, dry	3.6	2.9	2.9	2.9	3.0	3.1	3.0	3.0
Vegetable Oils, Oilseeds and Products (oil eq.)	11	14	16	19	20	21	18	18
Meat (carcass Weight)	63	74	80	80	87	91	75.6	76
Milk and Dairy, excl. butter (fresh milk eq.)	189	195	201	202	215	222	196	196

contd...

Kg/ Person/ Year	**1969/ 1971**	**1979/ 1981**	**1989/ 1991**	**2005/ 2007**	**2030**	**2050**	**Compa-rison 1991- 2001**	
							New	**Old**
Other Food (kcal/ person/ day)	492	508	498	458	488	509	472	458
Total Food (kcal/ person/ day)	**3138**	**3222**	**3288**	**3360**	**3430**	**3490**	**3251**	**3257**

Source: Alexandratos, N., and Bruinsma, J. *World Agriculture Towards 2030/2050*, ESA Working Paper No. 12-03, FAO, Rome. 2012, p. 44. Table 2.5.s.

On the other hand, the impacts of agriculture have so far been extremely harmful on the environment mainly through its expansion affecting 'habitats, biodiversity, carbon storage and soil conditions'. It has involved clearance of 70 per cent of the grassland, 50 per cent of the savanna, 45 per cent of the temperate deciduous forests and 27 per cent of the tropical forests with the main thrust in recent times more confined to replacing tropical forests (converting 80 per cent of its expansion estimated to release about 1.1×10^{15} grams of carbon per year, or about 12 per cent of total anthropogenic CO_2 emissions) while halting such expansion accounting for 98 per cent of total CO_2 emissions could be achieved along with saving the biodiversity and other ecosystem services.

Side by side, agricultural intensification with doubling irrigation during the last 50 years or so accompanied by about 70 per cent withdrawal of global freshwater, increasing fertilizer use by 500 per cent increasing crop-yields exceeding that if agricultural expansion has brought in its stride 'water degradation, increased energy use and the widespread pollution'. In addition, fertilizer use, manure application and the use of leguminous crops to fix nitrogen in the soil as a part of agricultural intensification had their severe impacts on 'water quality, aquatic ecosystem and the marine fisheries.'[10]

According to Tilman and Clark (2015), the relationship between food, agriculture and environment has become highly complicated through mismanagement by overzealous attempts by man making agriculture and food production responsible for more than 25 per cent (30-35 per cent) greenhouse gas emissions to the atmosphere. Application of fertilizers 'possessing more nitrogen and phosphorous than supplied by natural processes' which majority of the crops 'do not take up and leached out of farmed fields' they pollute aquifers, lakes and rivers in a process called '*eutrophication*', and these excess nutrients entering the ocean create low-oxygen *dead-zones* devastating 'local aquatic ecosystems and fisheries on which we thrive.[11]

Tilman and Clark (2015) are of the opinion that: 'If agricultural practices were to stay the same and the per capita food demand did not increase, population expansion (9.6 billion by 2050 and 10.9 billion by 2100) alone would increase demand for agricultural production by 30 per cent by 2050 and 50 per cent by 2100.'[12]

It may be noted that since rising income is a very strong driver for making dietary changes, demand for agricultural crops increases much faster than global population. Taking a stock of the situation it is estimated that the food that enters a typical household in the fifteen richest nations of the world amounting to about 3500 calories per person, '25 per cent of it is wasted'. Regarding composition '20 per cent of this food in the wealthy households consist of meat, milk, eggs'. In comparison, the average food demand in the twenty or so poorest economies is about 2000 calories consisting of only '3 per cent from meat, milk and eggs' without any wastage. The interesting thing is that the difference in calorie demand makes '2.1 billion people overweight or obese because of excess consumption' while another 800 million, the world's poorest people suffer from malnutrition without access to 'adequate and appropriate food'.

It should be noted here that the greater per capita consumption of animal products among richer nations makes a demand of 8000 calories per day of agricultural crops as animal feed.

> For example, for cattle to create one kilogram of edible beef protein they must eat 20 kilograms of plant protein, for a ratio of 1:20. The ratio is 1:5.7 for pork; 1:4.7 for poultry; 1:44 to 1:48 for aquaculture production of halibut, salmon, cod and arctic char; 1:3.9 for milk; and 1:2.6 for eggs. In total, more than half of all agricultural crop production and land use in richer nations feeds livestock, rather than producing crops for direct human consumption ... approximately 80 per cent of global agricultural land is used to graze livestock and a significant portion of cropland is for animal feeds.[13]

Thus, dietary shifts with rising income and growing preference for meat, poultry and other animal products has given rise to serious health problems, like increasing rates of obesity subjecting some 2.1 billion to obesity, heart diseases, diabetics etc., both in the developed and developing countries endangering human life. Global dietary transition is one of the greatest challenges facing humanity. For example, 'dietary composition strongly influences GHG emissions'. Animal-based foods emit more GHG emissions than plant-based foods with large differences like ruminant meats (beef and lamb) having emissions 'per gram of protein 250 times those of legumes.' It is important to note that ruminant meats have higher emissions per gram of protein than eggs, dairy, non-trawling seafood, aquaculture, poultry, and pork. The method of production of food as well affects emissions, for example seafood produced by nets dragged across the ocean floor emits 3 times more than that non-trawling seafood. Even items belonging to the same food group, like rice and wheat, vary in emission-levels, wheat having 'a fifth of GHG emissions per gram of protein of rice'. On the other hand, '20 serving of vegetables have less GHG emission than the one serving of beef.'[14]

At the same time, it is highly important to note the impact of diet on the environment with variable GHG emissions. Composition of diet, on the other hand, is ultimately associated with human health as a source of disease incident-rates irrespective of income-level. Such differences are related to human health as well as their environmental impacts. These diets are called, as mentioned earlier, Mediterranean, Pescetarian and Vegetarian. The difference in the composition of each diet is quite significant as 'grains, vegetables, fruits, sugars, oils, eggs and dairy' with not more than the consumption of meat or seafood once in a month characterises a vegetarian diet in sharp contrast to a pescetarian diet which is a vegetarian diet including seafood while the Mediterranean diet is rich in vegetables, fruits and seafood along with grains, sugars, oils, eggs with moderate amounts of meat. In contrast, the omnivorous diets include all food groups.

The most significant health effect observed in each diet (except omnivorous), Mediterranean, Pescetarian and Vegetarian diets is a quite striking rate of reduction in diseases, which is 16-41 per cent in the case of Type II Diabetes, 7-13 per cent in Cancer and 20-26 per cent in Coronary Heart Diseases.[15]

At the same time, GHG emissions have a close link with diet and it has been found that 'per capita reduction' in emissions would amount to 30 per cent, 45 per cent, and 55 per cent for Mediterranean, Pescetarian and Vegetarian diets, respectively. Tilman and Clark conclude:

> Changes towards healthier diets can have globally significant GHG benefits. From 2009 to 2050 global population is

> projected to increase by 36 per cent. When combined with the projected 32 per cent increase in per capita emissions from income-dependent global dietary shifts, the net effect is an estimated 80 per cent increase in global GHG emissions from food production.... This increase is equivalent to total 2010 global transportation emissions. In contrast, there would be no net increase in food production emissions if by 2050 the global diet had become the average of Mediterranean, Pescetarian and Vegetarian diets.[16]

It is quite natural that the composition of diet of income-dependent people will go through great changes in future, putting pressure on agricultural production responding to demands of the consumers.

IV. Impact of Agriculture on Environment

Expansion of Agriculture into other ecosystems causes very adverse effects on 'habitats, biodiversity, carbon storage and soil conditions'. We have noted from Foley *et al.* (2011) that agriculture has already created worldwide havoc by way of clearing or converting other ecosystems to the extent of about '70 per cent of grassland, 50 per cent of the savanna, 45 per cent of the temperate deciduous forests and 27 per cent of the tropical forests, particularly replacing tropical forests for 80 per cent of its expansion in the process of destroying rich reservoirs of biodiversity and accounting for about 98 per cent of the CO_2 emissions from land clearing. At the same time, agricultural intensification process during the past 50 years or so doubling irrigated areas, increasing fertilizer-use by '500 per cent' (over 800 per cent in nitrogen alone) has caused 'water degradation, increased energy use, and widespread pollution'. As mentioned earlier, use of fertilizer, manure application and leguminous crops for fixing nitrogen in the soil have disturbed global nitrogenous and phosphorous cycles besides enhancing GHG emissions with 'methane emissions from livestock and rice cultivation and nitrous oxide emissions from fertilized soils'. The authors clearly point out the severe harms inflicted on the environment so far through the practice of agriculture.

First, the expansion of agriculture in the tropics is reducing biodiversity, increasing greenhouse gas emissions, and depleting critical ecosystem services. Yet this expansion has done relatively little to add to global food supply; most production gains have been achieved through intensification. Second, the costs and benefits of agricultural intensification vary greatly, often depending on geographic conditions and agronomic practices. This suggests that some forms (and locations) of intensification are better than others at balancing food production and environmental protection.[17]

Hence, it emerges that agricultural policy needs to be carefully crafted so that it can take care of food-security as well as environmental sustainability.

According to Jonathan A. Foley *et al.* (2010), first of all the supply of food needs to be doubled in the next two decades, meaning to take the challenge of food security for all global population growth by 2050 (to 9 billion or so?). Along with this, the transformation or change which should be done to agricultural practices with the aim of cutting GHG emissions by at least 80 per cent, involves reducing losses of biodiversity and habitat, reducing unsustainable water withdrawal to meet competing demands, and reducing water pollution through phasing out the use of chemicals. In order to meet the dual challenges of agricultural practices for food security and environmental challenges certain steps have been suggested by these experts as mentioned below:

1. Agricultural Expansion has to be stopped to put it on a sustainable path.
2. Closing yield-gaps, 'increasing food production on existing agricultural land', in other words we have to increase yield-rates on the land currently under production which are not giving their best or, rather, 'underperforming', meaning the full potential of the land is not derived'.

For example, in general, yield-gaps are low in developed countries which apply to crops like maize, potato, rapeseed, rye and sunflower, in Western Europe as well as apply to maize and soyabean in the United States while there are high yield-gaps for wheat in the US and eastern Canadian Plains. Whenever high yield-gaps occur in Western Europe they are concentrated in southern countries like Spain, Portugal, and Italy. At the same time, high yield-gaps are more characteristic for most crops including maize, wheat, barley, rapeseed, and sunflower grown in Eastern Europe. It is also high for crops like potato and rye grown in Europe.

In comparison, the pattern of yield-gap seems to be more variable in Asia than in Europe and the United States. For example, 'cluster of low yield-gaps are associated with more populous provinces of China growing rice, wheat, millet, potato and rye as well as in 'some parts of the Indo-Gangetic Plain producing rice, wheat and rapeseed'. In contrast, yield-gaps are enormously high across Africa for many crops 'especially growing rice and maize. However, lower yield-gaps in Africa usually occur 'in a strip' extending across parts of Western and Central Africa producing cotton, millet, groundnut, and sorghum while potato and sunflower are grown across Southern Africa. Further, 'cassava and pulses grown almost throughout the continent experience low yield-gaps. The situation in South America in this regard is contrasting in having consistently low yield-gaps for soyabeans in Central and South America, often low for sugarcane but consistently high for maize in Mexico though it is low in Brazil and Argentina.[18]

The primary aim is to maximise production from the land already engaged in production and without any further land-expansion (at the cost of environment, to meet the growing need for food, feed, and biofuel, and by closing yield-gaps). Intensive agriculture does not appear to be the solution to increase the production-level since land ultimately becomes degraded and less productive. High-yielding farming practices devised by continued research provide the only alternative to increase production reducing yield-gaps.

Much of the world's croplands produce significantly less food than their potential which can be realised with the application of 'proper technology and management techniques'. This difference between existing production and the actual potential is called yield-gaps, thus making great difference between how much is produced and what could really be produced.

> Nations with lower per capita incomes tend to have larger yield-gaps. For instance, much of Sub-Saharan Africa has maize, wheat, and rice yields that are only 20-25 per cent of what could be attained in those nations using improved technologies and practices (yield-gaps of 75 per cent to 80 per cent). These yield-gaps suggest that improved agronomic methods could increase the food supplies by 300-400 per cent while using only the existing croplands of these nations. One reason for these immense yield-gaps is that the poorest nations currently apply less than five kilograms of nitrogen fertilizers per hectare (and similarly low amounts of other inputs) whereas the richest nations apply thirty times that amount.[19]

The same authors, David Tilman and Michael Clark (2015) point out that 'of the one hundred most populous nations, fifteen nations with incomes nearest the global median income have fertilization rates that are less than half of these fifteen richest nations' and have yield-rates 'one-third of the richest nations'. According to their calculations 'closing yield-gap should be a top priority and if all the nations achieve '95 per cent of their attainable levels (still leaving 5 per cent unattained), no new land would be needed to meet food demands for 2050. Besides, using fertilizers at rates of about 80 per cent of those in the richest countries would decrease environmental harm from over-fertilization but could increase food production by 70 per cent'.

On the other hand, at present the global pandemic of obesity, diabetes and heart disease resulting from nutrition transition has turned into a great life-threatening situation primarily for people having omnivorous diets. But already there are prospects of getting out of this by way of changing the composition of daily diet to one high in vegetables and fruits, nuts and seeds, whole grains, and fish, called Mediterranean diet. One can also turn to a vegetarian diet as a definite rescue from the difficult situation, both Mediterranean and vegetarian diets having 'between 15-40 per cent lower incidence of diabetes and 20-25 per cent slower mortality from coronary heart disease'. Similar benefits may also be derived from Pescetarian diets which are basically vegetarian diets including seafood in place of dairy and eggs. It may further be interesting to know that these alternative diets lead to 'significantly lower global emissions' besides decreasing land clearing between now and 2050' by about 500 million ha.[20]

Another very important factor related to food security is that almost 'one-third of global agricultural production is wasted. According to Smil (2002, Paperback), loss after the crop is harvested taking place at different phases – 'handling, threshing, drying, storage, and milling' may be 'well below 10 per cent to as much as 40 per cent' (even higher figures reported for some African crops).'[21] Along with this another very important feature is food-waste particularly in rich households which if it could be reduced by half, almost 300 million hectares of land could be saved from cultivation along with GHG emissions from crop production reduced by one-fifth. Another very surprising revelation about food-waste shows that in a year, mentioned earlier.

> barbecue food-waste-bill worth £428 million (was made in Britain), as reported by Rebecca Smitherns in *The Guardian*, UK, dated 17 August, 2017, with an average of £36.47 food (per person?) mostly consisting of 'salad leaves, burger rolls, hot dog buns and potato salad are thrown away at each event', the number of which held being 12 million in the month of August 2017 in this country.[22]

Tilman and Clark (2015), further argue that 'greater consumption of plant-based and preferential consumption of low feed-input animal-based foods such as eggs, milk, and poultry meat' along with 'elimination of food-based biofuels' make it possible to feed an extra one billion people without any agricultural expansion and at the same time reduce emissions of GHG. This along with the 'shrinking or closing yield-gaps, on the other hand could feed '70 per cent or more of the food demand anticipated by 2050' with production of the existing cropland (without any further expansion) while it 'could save global biodiversity' on the other hand making determined aim through proper investments to 'increase yields in developing nations.'[23]

In brief, four core strategies adopted with serious intentions may solve the present crisis of food security, stop further abuse of the environment which, in fact, sustains agricultural activities; First, further agricultural expansion has to be stopped outright; Second, crop yield-gaps have to be minimised (or eliminated, if possible); Third, agricultural resource efficiency has to be increased to make the process of intensification more beneficial in terms of water-use, chemicals, manure, leguminous crops, nutrients; and lastly, Fourth, there should be changed in dietary preferences and reduction or halting of food-wastes.

The twin challenge of food security and environmental problems properly handled will increase global food availability by '100-180 per cent' likely to meet the projected demand for human-food by 2050, besides 'lowering GHG emissions, biodiversity losses, water use and water pollution'.**

V. How to Tackle Food-Security and Keep the Environment Unmolested

Food consumed by man has a very crucial role in making harmful effects on the environment such as one of the most important, greenhouse gas emissions. Tilman and Clark (2014) give evidence that plant-based foods 'have lower GHG emissions compared to animal-based food.' For example, ruminant meats (beef and lamb) 'have emissions per gram that are about 250 times those of legumes' which also applies to 'eggs, dairy, non-trawling seafood, traditional (non-recirculating) aquaculture, poultry and pork having lower emissions compared with ruminant meats.

But global dietary change particularly associated with rising income in the developing countries is a greatly disturbing factor with concomitant increase in the consumption of meat (ruminant, particularly). According to Tilman and Clark (2014), there was 750 per cent greater demand for ruminant meat in 2009 in the richest nations. The same trend is prevalent among some of the developing countries including China though India makes an exception to this in a big way keeping to meatless diets.

However, the most notable features regarding overall food consumption in recent years is that between 1969/1971 and 1999/2001 there have been 'significant' improvements in per capita consumption rising almost 400 kcal per day, from 2411 to 2789 kcal indicating lowering the rate of undernourishment (John Kearney, 2010). But there are still some developing countries, especially in Sub-Saharan Africa, namely Somalia, Burundi, Rwanda, and Kenya where per capita food consumption which was already low has become lower meanwhile.

It is, further, interesting to know the role of individual foodstuffs and their consumption, as pointed out by John Kearney (2010) as given below.

(a) *Cereals* are, by far, the most important source of food in the world, providing 50 per cent of calories and almost 54 per cent in developing countries, which may be as high as 70 per cent in some parts of Asia and Africa, but showing some decline in the predominately rice-based diets in countries like China and East Asian countries. In contrast, cereals contribute about 30 per cent of energy intake in industrial countries showing signs of decline. In contrast to rice, global consumption of wheat, particularly in developing countries indicating greater prospect in future mainly in non-producing countries or in those where major diets are more dependent on 'roots, tubers, bananas and plantains'. In comparison, consumption of maize and millet, besides sorghum, is slowly declining, with the exception of North America where maize is used as a sweetener.

(b) *Meat* is 'centre-piece of most meals in developed countries' and in the 'USA and UK the most important sources are pigs, sheep and cattle' though of late chicken 'has become a major source in the UK while in the whole of Europe consumption of red meat has been decreasing faster than in the UK'. Interestingly, a major part of the increase in global meat consumption is attributed to poultry in contrast to fall in beef consumption in North America, Oceania and Europe but increase in China and Brazil. On the contrary while there has been a three-fold increase in meat consumption in the developing countries since 1963 with three-fold increase in Brazil and nine-fold increase in China, it is far from prominent in India and Africa.

(c) *Eggs, Milk, and other Dairy Products:* The consumption of eggs has doubled worldwide (per capita per day) more in the developing than in developed countries though both India and many Sub-

Saharan countries show little or no rise while both Brazil and China show a remarkable increase. In contrast, consumption of eggs shows modest rise in eastern Europe but modest decline in North America and very sharp decline in Oceania. In comparison intake of milk has risen in a number of developing countries, mainly in Asia but has declined sharply in the USA replaced by carbonated beverage and juices. On the whole, as Kearney shows, consumption of eggs will 'continue to rise' in contrast to milk-consumption falling continuously, when there will be no appreciable change in the consumption of butter and cheese.

(d) *Fish:* Worldwide increase in fish catches is depleting existing stock amounting to over-fishing. Popular varieties are white fish, oily fish, and seafood invertebrates. Fishes are source of 'good quality proteins and low in fat, besides being a major source of iodine.' But consumption of fish in the UK is low (22g per capita per day) compared with Portugal, Spain, and many other European countries. According to the report of John Kearney (2010), globally there has been 'little or no increase in the consumption of demersal, marine or pelagic fishes while consumption of sea food and freshwater fishes has experienced appreciable increase since the early 1960s. Both Oceania and Asia, especially China, has recorded highest interest in sea food, increasing about '11g per capita per day in 1963 to about 69g per capita per day in 2003 – a ten-fold increase.' On the whole consumption of sea food is 'set to continue to rise towards 2050' at a faster rate than other categories of fish.

(e) *Vegetables, Roots, Tubers, Pulses and Fruit:* The consumption of roots and tubers consisting of cassava, sweet potatoes, potatoes, yams, faro, and plantain show very modest declines worldwide 'particularly in China and Sub-Saharan Africa'. However, the major contrasting picture shows 'sharp fall in the consumption of sweet potatoes in many developing countries against sharp rise in consumption of potatoes particularly in a number of developing countries, China exhibits contrasting levels of consumption between sweet potatoes dropping '227g per capita per day in 1963' to 99g in 2013 when consumption of potatoes increased from 25g to 96g per capita per day during the same period. The consumption of pulses has declined globally, particularly in the developing countries while inadequate consumption of fruits and vegetables, production of which has been increasing over recent years, requires market supply-chains need to be devised for making fruits and vegetables more accessible for poor households (John Kearney, 2010).

(f) *Energy Providers: Vegetable Oils, Animal Fats and Sugar:* Among the energy providers consumption of vegetable oils has recorded an impressive increase all-over the world 'threefold in developing countries and twofold in industrial countries, most marked in China, Brazil and India' thus increasing food-security in these countries and involving a drastic changeover from animal fats to vegetable fats and this trend is likely to continue in the future. Sugar, as another energy provider, has increasing consumption in developing countries, particularly in 'Asia, India and to a lesser extent in Latin America' but slowing down in 'some industrial regions like North America and Oceania but with a little rise in Eastern Europe.'[24]

In this connection, there are some very important factors taking an active role affecting nutrition transition which are called 'Drivers'. The most important of them without doubt is income particularly in developing countries and rising income according to John Kearney (2010) means 'higher fat diets' consisting of animal-based foods and processed foods. In contrast, however, we find evidence, such as increased income proving beneficial, 'resulting in better quality diet, better healthcare, lower morbidity and mortality from infectious diseases and lower risk of obesity'.

The next important driving force is urbanisation acting as the most characteristic force of population movement leading to 66 per cent of global population as urban-dwellers by mid-twenty first century and having greater effect in developing countries; bringing in its wake devastating changes in food habits increasingly dominated by supermarkets controlled by corporate giants with greater thrust on importing food from outside introducing fast changes in the dietary habits of people with fast-food facilities. In terms of nutrition henceforth, the urban population will have greater preference for processed/semi-processed food items like hamburgers, pizzas and fried chicken consisting of a diet with more fats and more animal protein from meat and dairy products. In fact, this transition is going to be a greater problem for developing than developed countries in future.

The third important driving force changing food patterns are big trade corporations controlled by large corporates. 'Reduction in the price of facilitating growing change in urban food habits containing calories-rich, nutrient poor and high in saturated fats and salt' are becoming more popular in developing countries through trade liberalisation policies. These big trade corporations include Transnational Food Corporations like KFC, McDonalds, Kraft and Nestle making available calorie-rich processed food cheaper than food made at home, and generally tastier and easier to handle than home-made food.

The fourth player driving changes in human food are the supermarkets of giant mercantile houses which have sprung up in the urban areas of the world and as John Kearney (2010) observes, in 2000 supermarkets have sold about 60 per cent on average of the national retail trade in South America and Mexico and such expansion is 'only about 5-7 years behind in East and South-East Asia as well as in Eastern and Central Europe' affecting the entire food retailing system 'from farm to fork'. However, there are some real benefits derived from supermarket system, such as they solved the problem of 'keeping animal-based products chilled' at competitive prices and cheap and most importantly making 'safe milk' available for the poor in Brazil. But at the same time, the very unwelcome side of the development of supermarket system is making available 'cheaper less healthy processed, higher fat, added sugar and salt-laden foods, especially in developing countries'.

Fifth, perhaps one of the most damaging drivers for human-health of the food-marketing system is the introduction of changes in food-consumption patterns, most health-damaging of which includes introduction of beverages. For example, according to John Kearney (2010) in the USA alone, where in 1945 'Americans drank four times more milk than carbonated soft beverage drinks, 50 years later, they were consuming nearly two and a half times more carbonated beverages than milk'. Advertising and subsidies too have taken a very important role in popularising their sales.

VI. Nutrition Transition and Global Food Demand

As commented by Adam Drewnowski *et al.* (1997) 'both wealth and poverty have profound effect on diets, nutrition and health'. We have already observed how dietary changes are closely connected with income and through nutrition transition determine human health. The interrelationship between income and diet-structure ultimately determines the demand for food in its various forms production of which keep growing with global population in mind which finally makes agricultural policy and how to attain it.

The close bondage between high income associated with 'greater percentage of energy received from sugars and vegetables and animal fats along with sharply diminishing proportion of carbohydrates is the most characterising feature of nutrition transition and as we have already noticed, growing rate of urbanisation has a very important part in setting the course of this transition, or, rather in 'accelerating nutrition transition', According to Drewnowski *et al.* (1997), the intimate relationship between income and animal fats when 'poorest nations derived as little as 5 per cent of energy from animal fats in 1962'. However, a dramatic change took place by around 1990 since when the proportion of energy from vegetable fat was higher than animal fats and was not related to income. During the past five decades following the World War II there had been:

> a revolution in the production and processing of oil-seed based fats. Initially, a rapid increase in the world demand for milk and meat products after World War II fostered the demand for protein supplements. Although the world need for protein received the most research and popular attention, growing availability of cheap fats may turn out to have had more impact on the human diet. Technological breakthroughs in the development of high-yield oilseeds and in the refining of high-quality vegetable oils greatly reduced the cost of baking and frying fats, margarine, butter-like spreads, and salad and cooking oils in relation to animal-based products. Worldwide demand for fats was fuelled by health concerns regarding the consumption of animal fats and cholesterol. Furthermore, a number of major economic and political initiatives led to the development of oil crops not only in Europe and the United States but also in Southeast Asia (palm oils) and in Brazil and Argentina (soyabean oils). The net effect was that from 1945 to 1965, there was an almost fourfold increase in the US production of vegetable oils, whereas animal fat production increased by only 11 per cent.[25]

In consequence further initiatives were taken in Europe and North America leading to the fourfold increase in the production of vegetable oils when animal fat production increased by only 11 per cent between 1948 and 1965. 'Between 1991 and 1996/7 global production of 'vegetable fats and oil rose from 60 million to 71 million metric tons' in comparison to the production of butter and tallow remaining steady at 'about 12 million metric tons'. The major sources of vegetable oils are 'soyabeans, sunflower, rapeseed, palm, and peanut among which soyabeans now account for the bulk of vegetable oil consumption worldwide'. Further, it is very interesting to observe in this connection as pointed out by authors, the difference in dietary trends between rural and urban areas such as consumption in urban areas is characterised by 'superior, or polished grains like rice, wheat rather than corn or millet, more fats and animal products, more sugar and more processed foods, and more foods are taken away from home' (in restaurants and the like). Further, urban food is made tastier with greater addition of sweetmeats.

The picture of nutrition-transition becomes clearer from a study of regional variations. For this purpose, we may take up the cases of China and Japan having contrasting food habits.

The nutrition-transition in China has a close link with 'rapid urbanisation, a decrease in fertility, decline in mortality and increase in health and education'. Having attained a high degree of food security, it had remarkable change in the character of the diet, shifting very fast from traditional Chinese diet very low in fat, a carbohydrate-based diet (very similar to the diet of Indian rural population even now) indicative of poverty in the country. With rising income, the carbohydrate meals were abandoned, replaced by a diet with over 30 per cent energy from fat, and between 1991 and 1996 China's oilseeds production increased by 50 per cent along with imports of vegetables having tripled while imports of soyabean oil 'alone increased by 15 times (from 0.1 million to 1.5 million tons).

Both Drewnowski and Popkin (1997) believe that 'diversity and variety add to the essential component of eating pleasure' and with growing income diets become 'more diverse' since they contain a much larger variety of items like 'meat, fish, milk, eggs and cheese as well as vegetables, fruits' and sweets to make them more palatable. However, a study made in 1989 categorised consumption of Chinese-diets showing nutrition-transition linked to three income-groups. The diets of the poor were largely based on 'rice, millet, sorghums, cabbage, salted vegetables, soyabean sauce, and salt, with some meat (pork). In comparison, the higher income-groups 'replaced coarse grains and starchy tubers with more rice and wheat.' With more income 'vegetables and pickled vegetables were replaced with more meat, poultry, eggs, dairy products and fresh fruit.' With the use of a greater variety of food, protein of each item slowly decreased in size except for 'meat, fresh fruit and alcohol'. The notable feature of Chinese food is conspicuous by prevalence of high plant-food which is indicative of 'poverty rather than preferences'.

On the other hand, Japanese diets are rich in fats and sugar. However, the diet traditionally has been dominated by cereals like rice some 66 per cent of 'daily energy' which, however, has declined to 39 per cent, while in contrast to earlier patterns, the consumption of meat, mostly chicken

(35 fold) with milk as a major consumption increasing 'six-fold' and eggs (five-fold)'. Fish consumption is also high but has changed from the cheap varieties like sardines, mackerel, and herring to expensive varieties such as, 'salmon, tuna, shrimp and lobster'. At the same time, consumption of 'vegetable fats and oils increased from 6 to 31.5 pounds while fruit consumption has rather declined' since 1970. But what is important to note is all changes in diets were based mainly on domestic production rather than imports from outside, except meat and poultry done in significant quantities.[26]

VII. Global Trend in Food Consumption

The emerging global dietary patterns are characterised by 'a faster dietary diversity and increased proportion of vegetable fats and, to lesser extent, sugars'. Interestingly, increasing intake of vegetable fats as substitute for animal fats no longer makes income a controlling factor. As a matter of fact, nutrition transition has been responsible for greater variety in diet along with increased consumption of sugars and fats. It is highly doubtful if any control can be imposed on dietary changes which it seems to be left to its evolutionary process.

Since we are more concerned about demand for food keeping in step with rising population and finding out where it will come from, it should be noted that in 2000 'per capita consumption of calories and protein by the richest nations were 256 per cent and 430 per cent greater, respectively, than consumption by poorest nations.'[27]

Global crop products need to be increased by 100-110 per cent (Popkin, 2017) or 70-100 per cent (Godfray *et al.*, 2010), according to two different sources, which needs to be examined further. As pointed out by David Tilman *et al.* (2011) future production of crops can be increased 'either by way of extensification, this is agricultural expansion bringing more land under crop production or through intensification achieved with higher yields through increased inputs, improved agronomic practices, improved crop varieties and other innovations.

According to them:

> agricultural intensification through technology adaptation and transfer and enhancement of soil fertility in poorer nations would greatly reduce these yield-gaps, provide a more equitable food supply, and greatly decrease the GHG emissions and species extinctions than otherwise would have resulted from land clearing. Our analyses explore the implications of 100 per cent increase from 2005 to 2050 in global crop

> production that we forecast. If global crop production demand were lower, less land clearing and/or use would be needed, and environmental impacts would be smaller.[28]

Before going any further, it is necessary to consider what sort of action needs to be taken to properly handle the emerging issue of global population which is expected to reach almost 9 billion by 2050. For a clear idea about the present global food system, it is important to look at the evolving nutrition transition. According to Barry M. Popkin (2017), there has been revolutionary changes in the food chain. Through his research investigation, Popkin (2017) has found that both in China (2011) and Mexico (2012), a large proportion of the food consumed is processed and packaged. Further investigations in the United States has also revealed that packaged and processed food dominates the food industry. Popkin is of the opinion that the food chain in India, Bangladesh and China is controlled by food manufacturers, retailers and food companies in high-income countries like the United States of America.

The result of the change in the food system on such a massive scale has made 2 billion people in the world overweight or obese, facing higher-mortality risks, making the necessity of changes in the diet system more urgent. Most frustrating thing is that supermarkets and mega markets like Tesco in Britain, Aldi, Walmart, Carrefour have taken over European Marketing Systems and such changes more recently have happened in Asia in three food-purchasing transformations, beginning first in Korea and Japan followed by second in Indonesia, Malaysia, the Philippines and Thailand and lastly reaching China and India.[29]

The growing popularity of processed food may be ascertained from the fact that in Africa the practice moved from South Africa chains to both sides of the continent while another study showed that '30 per cent of all food' consumed in China in 2012 'came from packaged and processed food' while in Mexico 'a much larger proportion of packaged and processed food' was consumed in 2011.

According to Popkin (2017), global food shows an increase in the 'levels of consumption from non-essential foods containing excessive added sugars, added salt, refined carbohydrates, SSBs (or sugar-sweetened beverages), grain-based deserts, and snacks'. In his opinion, 'two major shifts in diet' playing an important role concerned with global sustainability, (a) shifts towards increasing consumption of animal-source food in Lower-Middle-Income countries while there has yet been no effective reduction of these products in higher-income countries, and (b) where the consumption of ultra-processed food has increased.

On the other hand, poor infant-feeding practices, being fed large proportion of SSBs, savoury snacks, and many other types of highly processed nutrient-poor foods have compelled many governments into prohibitory actions, such as (1) taxation of SSBs and other broader categories of non-essential junk foods, (2) marketing control with greater part of it focussed on children, (3) front-of-package labelling profiles with a positive or negative logo, and (4) special regulations related to schools and/or other government facilities.[30]

It is very clearly pointed out by Popkin (2017) that what we really need is to get over the problem of unsatisfactory food regime as prevalent now irrespective of Global North or Global South, keeping in mind the fact that a sustainable diet means to have greatly reduced portions of beef and milk, and also perhaps with great reduction in having packaged, processed or semi-processed food, with greater preference for home-cooked food which is the only way to restrain use of sugar, salt and even vegetable fats along with a healthy combination of food items.

However, the problem with the human society is that in spite of the sign of deceleration the global population is likely to stabilise, or plateau at 9 billion by, or in the middle of the present century, which will create a great pressure on the food system with higher purchasing power meaning higher consumption and a 'greater demand for processed food, meat, dairy, and fish'. As a result according to Godfray *et al.* (2010), the world is faced with great challenges like rapidly changing demand for food for larger and more affluent population without increasing environmental footprint and affecting 'social sustainability and ensuring the poorest people are no longer hungry'.

> This challenge requires changes in the way food is produced, stored, processed, distributed and accessed that are as radical as those that occurred during the 18th and 19th century Industrial and Agricultural Revolution and the 20th century Green Revolution. Increase in production will have an important part to play, but they will be constrained as never before by the finite resources provided by Earth's lands, oceans and atmosphere.[31]

As has already been mentioned earlier, the scope of increasing grain production doubled during the last five decades with agricultural expansion of about 9 per cent, which is not encouraged in view of protecting diversity, carbon storage in rainforest given higher priority when much of agricultural land has already been lost to urbanisation and other human use as well as desertification, salinisation, soil erosion and other consequences of unsustainable land use besides further losses caused by climate change

(Godfray *et al.*, 2010) along with increasing use of food crops for the production of biofuel. All these naturally create a compulsion to produce food from the same amount, or even less amount, of land available.

Recent studies suggest that the world will need '70 to 100 per cent more food by 2050'. The next question arises as to how to get to the target? Meeting the gap-in-yield has been suggested as one of the major steps towards increasing production. Perhaps, the best bet for increasing yields depends on farmers' capability to access 'current genetic material and available technologies and management and to access and use, among other things seeds, water, nutrients, pest management, soils, biodiversity and knowledge'. It has been estimated that average climate-adjusted rice yields are 2.5 metric tonnes per hectare where irrigation is available in South-east Asia, yet the average yields achieved are 60 per cent of this figure Similar yield-gaps are found in rainfed wheat in Central Asia and rainfed cereals in Argentina and Brazil.[32] 'Substantially more food, as well as income to purchase food, could be produced with current crops, and livestock if methods were found to close the yield-gaps.'[33]

One of the major handicaps in increasing food production is the lack of proper storage and transport facilities to reach the consumers' markets. Besides, food production in developing countries may be severely affected by market intervention in developed countries taking in the form of 'subsidies or price supports'. According to Godfray *et al.* (2010), 'closing yield-gap would dramatically increase supply of food but with uncertain environmental impacts like greater emission of methane and nitric oxide more damaging than CO_2 emissions ultimately acting against potential food supply. On the other hand, agricultural sustainability implies 'use of resource at rates that do not exceed earth's capacity to replace them' so that 'dependency on non-renewable inputs is unsustainable'. In their opinion, promoting sustainability does not reduce yields or profits as has been experienced from one of the sustainability projects in developing countries, mentioned earlier in this chapter involving 286 agricultural sustainability projects in the developing countries, involving 12.6 million small holder farmers on 37 million hectares experienced an 'average yield increase of 79 per cent across a very wide variety of systems and crop types'.

But the authors point out here that 'lack of secure land-rights' may ultimately act as 'disincentive' for small landholders in working hard for getting to the right yield capacity of the land. They are of the opinion that only if such rights and welfare of the producers – the small landholders – get the right treatment and are guaranteed about external investments that will build up proper trust in the investors to bring about improvements in crop production and processing. Besides, liberalisation of world trade will stimulate overall production in developing countries.

Another point raised and discussed earlier is food-waste in both the developed and developing countries ranging between 30 per cent to 40 per cent of the produce. In India, for example, almost 35 to 40 per cent is wasted in the absence of cold storage facilities. In Southeast Asia rice grain, easy to store, is wasted through pests and spoilage. However, it is very interesting to note that diet-change could greatly solve food problem if more people turn into vegetarians, having a well-balanced diet rich in grains and other vegetable products instead of meat and dairy products when almost one-third of global cereal production is fed to animals for meat. Such a change will not only deal with obesity but cut undernutrition at the same time.

Simultaneously, expansion of aquaculture taken seriously may provide nearly 3 billion people with '15 per cent of their animal protein intake'. Development in 'hatchery systems, feeds and feed-delivery systems and disease management' could go a long way in providing a permanent source of protein to a major section of global population. On the reverse, however, aquaculture 'may cause harm to environment' because of the release of organic affluents or disease treatment chemicals into the water they thrive in which, naturally, does not offer a safe system.

As clearly hinted by Godfray *et al.* (2010) sustainably feeding 9 billion people has no easy solution particularly in the face of fast change in the consumption of rich countries. 'The goal is no longer simply to maximise productivity, but to optimise across a far more complex landscape of production, environmental and social justice outcomes.'[34]

NOTES

***Foley, J.A.** *et al.*, *Solutions for a Cultivated Planet.* Nature. 2011, 478, p. 337.
'Increasing population and consumption are placing unprecedented demands on agriculture and natural resources. Today, approximately a billion people are chronically malnourished while agricultural systems are concurrently degrading land, water, biodiversity, and climate on a global scale. To meet the world's future food security and sustainability needs, food production must grow substantially while at the same time agriculture's environmental footprint must shrink dramatically. Here we analyse solutions to this dilemma, showing that tremendous progress could be made by halting agricultural expansion, closing 'yield-gaps' on underperforming land, increasing cropping efficiency, shifting diets, and reducing wastes. Together these strategies could double food production while greatly reducing the environmental impacts of agriculture'.

**Foley, J.A. *et al.*, *Solutions for a Cultivated Planet.* Nature. 2011, 478, p. 341.
Today, humans are farming more of the planet than ever, with higher resource intensity and staggering environmental impacts, while diverting and increasing

fraction of crops to animals, biofuels and other non-food uses. This must not continue; the requirements of current and future generations demand that we transform agriculture to meet the challenges of food security and environmental sustainability.

Our analysis demonstrates that four core strategies can-in-principle – meet future food production needs and environmental challenges if deployed simultaneously. Adding them together they increase global food availability by 100-180 per cent, meeting projected demands while lowering greenhouse gas emissions, bio-diversity losses, water use and water pollution. However, all four strategies are needed to meet our global food production and environmental goals, no single strategy is sufficient.

We have described general approaches to solving global agricultural challenges, but much work remains to translate them into action. Specific land use, agricultural and food system tactics must be developed and deployed. Fortunately, many such tactics already exist, including precision agriculture, drip irrigation, organic soil remedies, buffer strips and wetland restoration, new crop varieties that reduce needs for water and fertilizer, perennial grains, and tree cropping systems, and paying farmers for environmental services. However, deploying these tactics effectively around the world requires numerous economic and governance challenges to be overcome. For example, reforming global trade policies, including eliminating our strategies and tariffs, will be vital to achieving our strategies … The search for agricultural solutions should remain technology-neutral'.

1. Monbiot, George. "We can go on eating like this." *The Guardian*, U.K., 11 December, 2017, p. 25.
2. Alexandratos, N., Brunisma, J. *World Agriculture towards 2030/2050*. ESA Working Paper, Rome. 2012 (Revised); 12(03), p. 1.
3. *Ibid.*, p. 7.
4. Clark, Wilson. *Energy for Survival: The Alternative is Extinction*. 1973. Bombay: Allied Publishers Limited, 1977, p. 16.
5. Bhattacharya, Bimalendu. *Food: The Growing Challenge for Human Survival*. New Delhi: Concept Publishing Company Pvt. Ltd., 2020, pp. 118-121. (Tables 3.8 and 3.9).
6. Alexandratos, N, Brunisma, J. *World Agriculture towards 2030/2050*. ESA Working Paper, Rome. 2012 (Revised); 12(23), p. 17.
7. *Ibid.*, pp. 29-30.
8. *Ibid.*, pp. 45-46.
9. Foley, J.A., Ramamurthi, N., Brauman, K.A., Cassidy, E.S., Gerber, J.S., Johnston, M., Mueller, N.D., O'Connell, C., Ray, D.K., West, P.C., Balzer, C., Bennett, E.M., Carpenter, S.R., Hill, J., Moufreda, C., Polasky, S., Rockstrõm, J., Sheehan, J., Tilman, D., Zachs, D.P.M. "Solutions for a Cultivated Planet." *Nature*, 478, 2011, p. 337.
10. *Ibid.*, p. 338.

11. Tilman, D., Clark, M. "Food, Agriculture and Environment: Can we feed to the world and save the Earth?", *The American Academy of Arts and Science,* 144.4, 2015, pp. 8-9.
12. *Ibid.,* p. 9.
13. *Ibid.,* p. 10.
14. Tilman, D., Clark, M. "Global Diets Link Environmental Sustainability and Human Health." *Nature,* 515, 2014, pp. 518-519.
15. *Ibid.,* p. 520.
16. *Ibid.,*
17. Foley, J.A. *et al.* "Solutions for the Cultivated Plant." *Nature,* 478, 2011, p. 338.
18. Licker, R., Johnston, M., Foley, J.A., Barford, C., Kucharik, C.J., Monfreda, C., Ramamkutty, N. "Mind the Gap: how do climate and agricultural management explain the yield gap of grasslands around the world?". *Global Ecology and Biogeography,* 19.6, 2010, pp. 776-778.
19. Tilman, David, Clark, Michael. "Food, Agriculture and Environment: Can we feed to the world and save the Earth?", *The America Academy of Arts and Science,* 144.4, 2015, pp. 14-15.
20. *Ibid.,* pp. 15-16.
21. Smil, Vaclav. *Feeding the World: A Challenge for the Twenty-First Century.* Cambridge, Massachusetts: The MIT Press, 2002, p. 186.
22. Bhattacharya, Bimalendu. *Food: The Growing Challenge for Human Survival.* New Delhi: Concept Publishing Company Pvt. Ltd., 2020, pp. 244-245.
23. Tilman, D., Clark, M. "Food, Agriculture and Environment: Can we feed to the world and save the Earth?" *The American Academy of Arts and Science,* 144.4, 2015, pp. 17-18.
24. Kearny, John. "Food Consumption Trends and Drivers." *Philosophical Transactions of the Royal Society Biological Sciences,* 365.1554, 2010, pp. 2794-2798.
25. Drewnowski Adam, Popkin Barry M. "The Nutrition Transition: New Trends in the Global Diet." *Nutrition Reviews,* 55.2, 1997, p. 34.
26. *Ibid.,* pp. 38-39.
27. Tilman, D., Balzer C., Hill J., Befort, B.L. "Global Food Demand and the Sustainable Intensification of Agriculture." *Proceeding of the National Academy of Sciences of the United States of America,* 108.50, 2011, p. 20260.
28. *Ibid.,* p. 20264.
29 Popkin, Barry M. "Relationship Between Shifts in Food System Dynamics and Acceleration of the Global Nutrition Transition." *Nutrition Reviews,* 75.2, 2017, pp. 73-74.
30. *Ibid.,* p. 78.
31. Godfray, H., Charles, J., Beddington, J., Crute, I. R., Hadad, L., Lawrence, D., Muire, J. F., Pretty, J., Robinson, S., Thomas, S. M., Toulmin, C. "Food Security: The Challenge of Feeding 9 Billion People." *Science: New Series,* 327.5967, 2010, p. 812.
32. *Ibid.,* p. 813.
33. *Ibid.*
34. *Ibid.,* p. 817.

6
The Role and Responsibilities of Small Holder Farmers in Agricultural Revolution

Agricultural production has more than tripled between 1960 and 2015 partly through the introduction of new technologies called Green Revolution and also through agricultural expansion at the cost of tropical forests. However, this will not eradicate global hunger even by 2050 if patterns of consumption in the form of nutrition transition are not tackled properly taking care at the same time about human-health system along with GHG emissions and biodiversity loss. The question that arises here is about the capability of agriculture to meet the unprecedented demand for food while keeping the natural resource base sustainable. Can agricultural sectors and rural economies be transformed providing more and better employment and income-earning opportunities in rural areas and stem mass migration to urban areas? Can the problem of food losses and waste amounting to a third of the production be tackled? These are some of the questions raised in the FAO Report (2017).[1]

I. Major Avenues of Increasing Agricultural Production

A great impediment to realising food security and reducing hunger comes from the fact that about 33 per cent of the world's farmland is moderately to highly degraded. On the other hand, agricultural expansion has stopped over the last 20 years (FAO, 2017) at around 4.9 billion hectares with forests (subtropical and tropical) amounting to a little less than 100 million ha. As a matter of fact, commercial agriculture has a very important role in deforestation in Latin America, and accounts for nearly two-thirds of the total forested land in Africa. The encouraging feature, as reported by FAO (2017) is that about 22 countries have reduced the prevalence of undernourished people besides improving agricultural production and food security while maintaining or increasing their forest cover in the past 25 years.

According to the same report, FAO (2017), growing water scarcity is going to put heavy stress on agricultural production in the face of stiff competition from industry and urbanisation. It is estimated that countries may be considered water-stressed if more than 25 per cent of their water, renewable freshwater, is used for agriculture. They approach physical water scarcity when more than 60 per cent is withdrawn and face severe water scarcity when more than 75 per cent is withdrawn. Such a situation along with the effect of climate change associated with heavy flood and droughts will have heavy impacts particularly on 'rainfed small holder farming in highland areas, and in the tropics, which account for 80 per cent of world's cropland and produce about 60 per cent of global agricultural output.[2]

It is highly interesting to note in this connection that though food and agriculture sector is one of the largest emitters of greenhouse gases, mainly caused by land use, livestock production, soil and nutrient management amounting to about 26 per cent of the total GHG emissions in the world, this can be effectively reduced by 15-25 per cent and lower methane emission by 30-70 per cent without affecting yields by way of 'alternate wetting and drying which is a water saving practice reducing water use in rice cultivation without affecting yields.'[3]

As reported by FAO (2017), making a thorough study of 1090 cases involving crops like wheat, maize, rice and soyabeans grown under different climatic conditions, climate change will 'significantly reduce yields' particularly in the low and middle-income countries but will increase yield in high-income countries in temperate zones enjoying a longer growing season, as mentioned elsewhere.

According to this report (FAO, 2017), food insecurity will increase further unless efforts are made to adapt to climate change. Increasing events like 'heat waves, droughts, floods, tropical storms and wildfires will have greater damaging effects on agricultural production and food insecurity' besides affecting 'food quality, food access and utilisation'.

It has further been evidenced that in some crops like wheat, rice, and soyabeans, the concentrations of minerals can be up to 8 per cent lower than normal under elevated levels of CO_2, while protein and carbohydrates being higher, will increase the incidence of diseases, particularly water-borne diseases such as diarrhoea undermining the body's ability to utilise the nutrition in food.

Agronomic adaptations can improve crop-yields by 15 to 18 per cent, but specific social and environmental conditions will influence small holders' choice of adaptation measures since climate change has different effects, such as while the application of mineral fertilizer may generate higher yields under average climatic conditions it may bring lower yields when rainfall is highly variable or delayed on crop rotation producing lower yields under average climatic conditions and higher yields under high rainfall conditions. In fact, 'adoptation of sustainable land, water, fisheries, and forestry management practices by small holder agricultural producers will be crucial to adapt to climate change, eradicate global poverty and end hunger yielding significant productivity improvements' (FAO, 2017).

Another very important feature of small holder agriculture is that women farmers make up more than half of agricultural workforce playing a crucial role in the management of natural resources. In spite of that 'they are systematically disadvantaged compared to men, with fewer endowments and commitments and more limited access to information and services'.

According to FAO (2017), agriculture in 2050 will have to produce almost 50 per cent more food, feed, and biofuel than it did in 2012 for feeding 9.73 billion expected to be the global population by then. In its opinion, agricultural output in Sub-Saharan Africa and South Asia 'would need to more than double by 2050, to meet increased demand' while in the rest of the world it should be 'about one-third above current level as shown in Table 6.1. In this connection, it is mentioned (FAO, 2017) that there are similar experiences of increase in agricultural output between 1961 and 2011 when global agricultural production more than tripled, and quoting Alexandratos and Bruinsma (2012), it is further stated that in 'low-income countries, livestock production had been one of the fastest growing agricultural subsectors. Since the early 1970s, per capita consumption of milk, dairy products and vegetable oils had 'almost doubled while meat consumption had almost tripled'.

Table 6.1: Increase in Agricultural Production Required to Match Projected Demand 2005-2050 (%)

	2005/7	**2050**	**2005/7-2012**	**2013-2050**
World				
As projected in AT2050	100	159.6	14.8	44.8
With updated Population Projections (U.N.2015)	100	163.4	14.8	48.6

contd...

	2005/7	2050	2005/7-2012	2013-2050
Sub-Saharan Africa and South Asia				
As projected in AT2050	100	224.9	20.0	104.9
With updated Population Projections (U.N.2015)	100	232.4	20.0	112.4
Rest of the World				
As projected in AT2050	100	144.9	13.8	31.2
With updated Population Projections (U.N.2015)	100	147.9	13.8	34.2

Source: FAO, 2017, *Rome: Future of Food and Agriculture: Trends and Challenges.* p. 46, Table 6.1.

Note: *AT2050: World Agriculture Towards 2030/2050: The 2012 Revision*: Alexandratos and Bruinsma.

This also applies to fish consumption having more than doubled over the past five decades and aquaculture became the 'fastest growing food production industry'.

Table 6.2: The Average Crop Yields, 2001-2012 (tonnes/ha)

Country/Group	Wheat	Rice	Maize
Low-Income	1.82	3.3	1.54
Lower-Middle-Income	2.74	3.65	2.74
Upper-Middle-Income	2.67	5.28	4.41
High-Income	3.5	6.64	8.99
World	2.92	4.16	4.87

Source: FAO, 2017 (Following FAO 2014).

However, according to Alexandratos and Bruinsma (2012), cereal production should increase from 2.068 million tonnes in 2005-07 to 3.009 tonnes in 2050, a 43.4 per cent yield increase for cereals which would take the most important place in meeting the higher food requirements in around 2050. However, since the 1990s, there have been increases in the 'yields of maize and rice but wheat has slowed down annually to a little more than 1 per cent at the global level than it was in the 1960s, while yield growth has been lower than 1 per cent annually for soyabeans and sugarcane. Besides, as shown in Table 6.2, yield-rates for different crops are quite different between low- and high-income countries. In this connection it is quite important to observe that the yield-rate for low-income countries and middle-income countries is lower even than the world-average with the exception of the case of rice which is higher while it is much higher for each of the three crops in higher-income countries.

II. Resource-Conserving Practices as Boosters for Agricultural Productivity

As observed by FAO, 2017:

> 'The key to sustainable agricultural growth is more efficient use of land, labour, and other inputs through technological progress, social innovation and new business models. For agriculture and aquaculture to respond to future challenges, innovation will not only need to improve the efficiency with which inputs are turned into outputs, but also conserve scarce natural resources and reduce waste. The farming practices required to consume and make more efficient use of natural resources will differ according to local conditions and needs.'[4]

Adoption of conservation agriculture in recent years is based on reducing soil disturbance by minimum tillage 'maintaining a protective cover on the soil surface and cultivating a wide range of plant species' consisting of both 'annuals and perennials – including rotations, including trees, shrubs and crops'. Promoting integration into cropping systems of pulses and legumes helps building up and maintaining soil nitrogen levels.

Conservation agriculture has been adopted worldwide on about 117 million ha or about 8 per cent of total world cropland, the highest adoption levels, about 50 per cent of cropland are concentrated in Australia, Canada, and the Southern cone of South America. Though such adoption is low in Africa, Central Asia, and China, it is however increasing. Practice of conservation agriculture has adequately been proved to provide farmers a greater quality and quantity of production by decreasing reliance on chemical inputs with greater emphasis on ecosystem management.

According to FAO, 2017, a 'climate-smart agriculture aims at sustainability increasing food security' adapting and 'building resilience to climate change.' Other innovations which area part of this approach are

'Conservation agriculture, agroecology, agroforestry and the development of crop varieties more tolerant to pests and diseases, drought, waterlogging and salinity'. The practice has been taken up very seriously by more than 30 countries in Sub-Saharan Africa. The practices of climate-smart agriculture put emphasis on mixed crop-livestock systems and sustainable livestock production including the 'rotation of pasture and forage crops to enhance soil quality and reduce erosion and the use of livestock manure to maintain soil fertility as an integral part of this system.'[5]

However, all these developments in agricultural production have been possible with increasing expenditure on R&D which slowing down in the 1990s increased by an 'average of 3.1 per cent a year between 2000 and 2009 rising from US$25 billion to U$33.6 billion. Interestingly, China and India took a leading part in it 'accounting for close to half of the increase'. Among other countries Argentina, Brazil, Iran, Nigeria, and the Russian Federation took significant part in it, collectively accounting for one-fifth of the total global increasc. Similarly, annual R&D spending grew by 2.3 per cent annually among the low-income countries with Ethiopia, Kenya, Uganda, and the United Republic of Tanzania taking greater interest in it.

Interestingly, private investment in R&D accounted for about 21 per cent of total R&D expenditures in 2008. Though almost all private research takes place in high-income countries both China and India account for respectively 19 and 16 per cent of it (Table 6.3).

Table 6.3: Growth in Public Spending on Agricultural R&D (%)

	1960-70	1970-80	1980-90	1990-2000	2000-2009
Country Group					
Low-Income	6.0	4.1	3.1	1.0	2.3
Lower-Middle-Income	6.7	3.3	3.1	3.3	4.6
Lower-Middle-Income (exc. India)	5.4	4.1	2.0	0.8	4.1
Upper-Middle-Income	4.3	6.6	2.8	2.1	6.2

contd...

	1960-70	1970-80	1980-90	1990-2000	2000-2009
Upper Middle-Income (exc. China)	6.1	6.2	1.9	1.1	3.0
High Income	9.1	3.3	2.1	1.6	1.0
East Asia/ Pacific (inc. China)	2.4	5.6	4.6	3.4	8.7
China	-0.1	7.8	5.1	3.9	9.9
Latin America/ Caribbean	4.4	9.7	1.0	0.9	2.5
Middle East/North Africa	0.0	2.0	5.2	2.5	3.0
South (inc. India)	8.2	3.2	5.7	5.3	4.4
India	9.8	1.7	5.6	6.7	5.2
Sub-Saharan Africa	5.7	2.7	0.6	-0.5	4.0
World	7.6	4.1	2.4	1.9	3.1

Source: FAO, 2017, Table 6.3 p. 52.

III. Changing Food Systems and Sustainably Improved Agricultural Productivity to Meet Increasing Demand

Agriculture and food production increasingly serving urban and peri-urban markets has gradually turned into a new form of processed or semi-processed food, mentioned earlier, ready for consumption catering to almost 40 to 50 per cent people in upper middle-income countries, increasing to 72 to 75 per cent in high-income countries through supermarkets while in North America 93 per cent of consumers thrive on processed or semi-processed food made available through hypermarkets, large super markets or convenience stores.

The situation of food availability for the growing global population is further aggravated by food losses and waste, as mentioned earlier, which on its own hold back transition to environmentally sustainable food systems.

> Globally around one third of all food produced is lost or wasted along the food chain, from production to consumption. In a world where hundreds of millions of people go hungry, that is a stark indication of the inefficiency of the current food system. Food losses and waste often translate into economic losses for farmers and other stakeholders within the food value chain, and higher prices for consumers, both of which affect food insecurity by making food less accessible for vulnerable groups. Reducing food losses and waste would increase the supply of available food and strengthen global food security.'[6]*

It has been noticed that intensive use of inputs in crop production has a strong deteriorating effect on sustainable agricultural production. It is advised that further investment needs to be made specially on 'agriculture, fishery and forestry, and on research and development (R&D) which needs to be increased' particularly 'in and for low-income countries for promoting the adoption of sustainable production systems and practices, 'including integrated crop – livestock and aquaculture – crop systems, conservation agriculture, agroforestry, nutrition-sensitive agriculture, sustainable forest management and sustainable fisheries management'. As viewed by the experts (FAO, 2017) increase in food 'will have to come mainly from improvements in productivity and resource-use efficiency'.

Despite efforts at reducing poverty during the last three decades there are still about 2.1 billion people worldover living in poverty with '700 million of them in extreme poverty'. Most of them live in the rural areas with agriculture as their livelihood, their highest share of earning spent on food, leaving very little for meeting absolute necessities. As pointed out by FAO 2017 in its report 'agriculture plays an important role in pro-poor growth' based on 'increased productivity and profitability and linking farmers to markets', and providing 'efficient extension and agricultural advisory services'. Further required are 'education for farmers, diversion through training in non-farm income generating activities, support for job creation and adequate social protection mechanisms'.

It is now, necessary to review growth in the use of national resources in agricultural production to assess the chances of satisfying the food situation in 2050. As estimated by Alexandratos and Bruinsma (2012), 'the world's average daily calories availability could rise by 30 per cent in kcal per person' by 2050, an '11 per cent increase over its level in 2005/2007, which still leaves 'some 4 per cent (about 320 million) of the developing countries' people chronically undernourished. Table 6.4 gives a comprehensive idea about likely increases in population through the period 1961/63 to 2050 including calories supply and agricultural production.

In order to realise these goals global agricultural production has to increase by about 60 per cent between 2005/2007 and 2050. This would consist of 77 per cent increase in developing countries and 24 per cent in developed countries. Incremental quantities involved are considerable as for example, 'by 2050 annual cereal production would have to increase by 940 million tonnes (+46%) and meal production by almost 200 million tonnes (+76%) a considerable part of which would require a substantial portion of the increased cereal production some 443 million tonnes (or about 60%) would be required as animal-feeds and 23 per cent for biofuels. As a result, the share of 'livestock production (meat, milk and dairy products and eggs)' would increase globally from 36 per cent in 2005/2007 to 39 per cent in 2050 (from 30 to 35% in developing countries)' (Alexandratos and Bruinsma (2012).

Table 6.4: Increases in population, calorie supply and agricultural production

	Unit	**1961/63**	**2005/07**	**2030**	**2050**
World					
Population	million persons	3133	6569	8276	9111
Daily Energy Supply	kcal/ person/day	2231	2772	2960	3070
Total Production*	index/(2005/ 2007 = 100)	37	100	138	160
Cereals**	million tonnes	843	2068	2720	3049
Meat Production	million tonnes	72	258	374	455
Developing Countries					
Population	million persons	2140	5218	6839	7671
Daily Energy Supply	kcal/person/ day	1884	2619	2860	3000

contd...

	Unit	1961/63	2005/07	2030	2050
Total Production	index/ (2005/2007 = 100)	24	100	147	177
Cereals	million tonnes	353	1164	1572	1812
Meat Production	million tonnes	20	149	243	317
Developed Countries					
Population	million persons	1012	1351	1437	1439
Daily Energy Supply	kcal/person/ day	2983	3260	3430	3490
Total Production*	index/(2005/ 2007=100)	64	100	118	124
Cereals*	million tonnes	500	904	1148	1197
Meat Production	million tonnes	52	109	130	138

Source: Alexandratos N., Bruinsma J. *World Agriculture Towards 2030/2050.* FAO Table 4.1. The 2012 Revision. p. 94.

Note: *In value terms (2004/06 International Commodity Prices).

** Including rice in milled form.

It is interesting to see that almost '90 per cent increase in annual production' will take place in the developing world, representing their agricultural production from 67 per cent in 2005/2007 to 74 per cent in 2050. At the same time, it is necessary to bear in mind that 'bulk of the food produced would be consumed locally' and only 19 per cent of the global production would enter international trade.

IV. Crop Production

Growth of crop production is seen to be decelerating in all regions, particularly in developing countries and East Asia with sharply contrasting features of 'slow growth of cereals like rice in contrast to more rigorous growth of coarse grains and oilseeds used for feeding livestock' meant for the increasing demand for meat as adietary choice.

According to FAO 2017, '80 per cent of the projected growth in developing countries would come from intensification in the form of yield-increases (73%) and higher cropping intensities (6%)'. More interestingly intensification contributes to 90 per cent in South Asia and

100 per cent of increase in Near East/North Africa and East Asia. As a matter of fact, intensification means more intensive use of land world-over which is going to account for the contribution of '90 per cent of the growth of crop production' in view of decline in arable land persisting 'since the mid-1980s'. FAO 2017 reports that 'the bulk of wheat and rice (will be) produced in the land-scarce regions of Asia and the Near East/ North Africa while 'maize in Sub-Saharan Africa and Latin America will be the major crop where some countries have got land for further expansion, undoubtedly through removal of forest. However, increasing interest of many countries in the production of more coarse grains lies in using them as feeding crops to livestock for the supply of more meat. China, as pointed out by Alexandratos and Bruinsma (2012) provides the most important example in this respect where rice and wheat production show declining trends in contrast to maize production which is expected to grow by 37 per cent.[7]

In getting a clear idea about the availability of land with crop-production potential, FAOSTAT maintains that in 2005/2007 'about 12 per cent or more than 1.5 billion ha of the global land surface was used for crop production (arable land and land under permanent crops). In this connection Fisher G. *et al.*, 2011 recommended a fresh category of land suitable for cultivation, known as Global Agroecological Zones (GAEZ), and according to this classification:

> about one-third (34%) of the world's land surface, or 4.5 billion ha is estimated to be of prime (very suitable) or good (suitable or moderately suitable) quality for rain-fed agriculture (Table 6.5). Of this area, some 1.6 billion ha is already under cultivation. It is interesting to note that of this 1.6 billion ha, some 300 million ha (or 19%) of agricultural land is on areas that GAEZ considers only marginally suitable or even not suitable, at least for rain-fed agriculture. Such areas might have been made productive by applying irrigation (e.g. 32 million ha in desert areas). Excluding irrigated area some 220 million ha (or 17%) of rain-fed agriculture is apparently taking place on marginally and not suitable areas.[8] (Alexandratos, N. and Bruinsma, J.).

On the whole, the developing countries have about 2.9 billion ha of 'prime and good quality land' of which a quarter, 700 million ha or 24 per cent is in use for agriculture. At global level, the gross land balance amounts to 3.2 billion ha, 2.2 billion ha of which lies in developing countries. It needs to be pointed out in this connection that gross balance

Table 6.5: Land with Rainfed Crop-Production Potential

(World: million ha)

	Total	**Poten-tial**	**VS****	**S**	**MS**	**mS**	**vmS**	**NS**
Total land*	**13,295**	4495	1315	2187	993	1111	1627	6061
of which in agricultural use 1999-2001	**1559**	1260	442	616	201	120	104	75
of which rain-fed land	**1283**	1063	381	516	166	93	84	43
of which irrigated land	**276**	197	61	100	35	27	20	32
Gross balance of land with rain-fed potential	-	3236	873	1571	792	991	1523	-
under forest	**3736**	1601	453	854	293	342	530	1263
Strictly protected land***	**638**	107	30	50	27	39	59	432
Built-up land	**152**	116	42	61	14	12	10	15
Net Balance of land with rain-fed potential	-	1412	349	606	458	598	923	-

Source: Alexandratos and Bruinsma, 2012, Table 4.6; p. 104.

Note: (Taken from GAEZ v3.0 in Fischer *et al.*, 2011).

*Crops considered: cereals, roots and tubers, super crops, pulses, and oil-bearing crops.

Sustainability classes are defined according to attainable yields as a percentage of the maximum constraint-free yield as follows: **VS = Very Suitable, 80-100%; **S** = Suitable, 60-80%; **MS** = Moderately Suitable, 40-60%; **mS** = marginally Suitable, 20-40%; **vmS** = very marginally Suitable, 5-20%; **NS** = Not Suitable, <5%. Potential = Prime Land **(VS)** plus Good Land **(S + MS)**.

***Of Land at present not cultivated, under forest or built-up.

ignores all other types of land, say forest cover, the built-up area other than only long growing crops, while there are still 1.4 billion ha worldwide of prime and good quality not cropped, 960 million ha of this land lying in developing countries quite erratically distributed. 'Some 85 per cent of the remaining 960 million ha in developing countries is to be found in Sub-Saharan Africa (450 million ha) and Latin America (360 million ha). In addition, about half of the remaining land is concentrated in just seven countries, Brazil, Argentina, Sudan, China, the Democratic Republic of Congo, Angola and Mozambique'. On the other hand, no prime and good land is available in many countries in the Near East, North Africa, South Asia, and Central America or in the Caribbean.[9]

Such a situation ultimately leads to the possibility of further expansion of land in crop production given a situation in the not-too-distant future when land devoted to agricultural production today declines slowly in proportion to the population. The problem is further aggravated in the face of 'finite natural resources such as land, water and genetic potential' and the need to preserve them unmolested for the use of our future generations 'through environmentally responsible and sustainable use', as pointed out by The Brundtland Commission Report in 1987. But it has already been pointed out that any agricultural expansion does not apply to land-scarce-countries and regions like South Asia and Near East/North Africa but it still applies to the 'unused large tracts of land with varying degrees of potential in several countries, such as in Sub-Saharan Africa and Latin America with some in East Asia, though in cases severely handicapped by physical constraints, as mentioned earlier, it may give rise to serious environmental problems in the long run when these lands are put to agricultural practices.

Yet, as pointed out earlier, 'expansion of land under agriculture' continues to take place under the pressure of 'growing needs for food and employment with limited access to technology packages that could increase intensification of cultivation on land already in agricultural use'. It is also happening in countries like Brazil with abundant land resources profiting from the growth of demand for exports, besides agricultural expansion taking place to meet heavy demand for soyabeans (as animal feed) in South America and oil palm in South East Asia. A good part of the total increase in cultivated land in the developing countries has been made at the expense of forest lands. It may be interesting to note that worldwide expansion of arable land between 1961/1963 and 2005/2007 by 176 million ha involved two opposite trends: 'an increase of 230m ha in the developing countries and a decline of 54 million ha in the developed countries' resulting from sustained yield-growth counterbalanced with a declining demand for agricultural products leading to further decline in arable land to 608 and 586 million ha in 2030 and 2050.

The eventual decline in the amount of arable land as a direct consequence of slowdown in crop production along with increase in crop yields estimated to increase resulting from global arable land by some 70 million by 2050 from the base year 2005/2007, making an increase by almost 110 million ha in the developing countries with a decline of nearly 40 million ha in the developed countries. According to an estimate made by Bringezu *et al.* (2010), 2 to 5 million ha of global arable land is lost every year due to soil erosion while another estimate mentions the loss of arable land amounting to 3 million ha annually due to severe land degradation.

However, arable land in developing countries is projected to expand by 107 million ha over this period extending up to 2050 but the harvested area would amount to '130 million ha or some 14 per cent, due to increase in cropping intensities. Cropping intensities continue to rise through shorter fallow periods and more multiple cropping increasing further through irrigation. An increasing share of irrigated land in total agricultural land contributes to more multiple cropping.'[10]

V. The Role of Small Holder Farmers in Global Food Security

However, the consistent rise of urbanisation has not yet 'overcome the stubborn persistence of the peasantry'. Whether this is due to the fact that 'historically new family-labour farms continually replace those lost through industrialisation, or because of the world's rural poor hardly has any other alternative but farming:

> the absolute numbers of peasants and small holder farmers in the global south have remained remarkably stable over the last forty years. Small holders continue to provide significant amounts of the food in the Global South, as high as 90 per cent of all food productions in African Countries.[11]

According to a report by an Indian expert, the small landholders in India having on average of less than two hectares, representing 80 per cent of farmers, 'contribute more than 50 per cent of the total agricultural output' on '44 per cent agricultural land' and support the 'livelihood and food security of millions of people' in the country.[12]

The small holder famers collaborated globally through building up in 1996 the La Via Campesina (The Peasant Way) defining very clearly their aim of making 'food sovereignty' as the right of the small holder farmers producing the food have the right of democratic control over the entire food-system – 'from production and processing, to distribution, marketing and consumption'. For decades these farmers, their families, the rural communities around the world have resisted the destruction of native seeds to be replaced by genetic seeds produced by giant corporate houses, have worked hard to diversify their crops, protect their soil through indigenous methods, conserve their water and forests' applying different locally developed practices like no/or zero no-till cultivation, drip irrigation systems, change in fallowing principles etc facing stiff competition from industrial agriculture which use farming practices to serve altogether different purposes. For instance, Table 6.6 provides a very clear idea with just one example about the use of crops by industrial houses in lieu of using them as human feed.

Table 6.6: World Use of Crops for Biofuels

		2005	**2030**	**2050**
Cereals	million tonnes	65	182	182
Cereals	per cent of total use	3.2	6.7	6.1
Vegetable Oils	million tonnes	7	29	29
Vegetable Oils	per cent of total use	4.8	12.6	10.3
Sugar (equivalent of sugarcane)	million tonnes	28	81	81
Sugar	per cent of total use	15.1	27.4	24.3
Cassava (fresh)	million tonnes	1	8	8
Cassava	per cent of total use	0.4	2.3	1.8

Source: Alexandratos N., Bruinsma J. *World Agriculture Towards 2030/2050*. 2012 Edition. Table A. 3.1, p. 92.

The same source[13] informs us that 742 million tonnes, or 36 per cent of world's total production of cereals and oil seed cakes are used for animal feed to increase meat for human consumption, no doubt, but it does not help in conquering hunger for malnourished people. According to these sources, per capita meat consumption in the developed countries will increase 14 per cent in the four decades to 2050.

Meanwhile, there had been revolutionary changes in the activities of the small holder farmers all over the developing countries, they are:

> working with movements like *Campesino a Campesino* (Farmer to Farmer) of Latin America and NGO networks for farmer-led sustainable agriculture like Participatory Land Use Management (PELUM) of Africa, and the Farmer Field Schools of Asia have restored exhausted soils, raised yields, and preserved the environment using highly effective agroecological management practices on hundreds of thousands of acres of land. These practices have given important measures of autonomy in relation to the industrial agri-food system and have increased their environmental and economic resistance buffering them from climate-induced hazards and market volatility.[14]

In consequence of the agreement made at Mons in Belgium in 1993 in a gathering of farm leaders coming from all-over the world 'the right of small farmers to make a living in the Countryside' was accepted along with their right to healthy food and the 'right of nations to define their own agricultural policies'. These agricultural groups making a living in the Countryside consisted entirely of marginalised groups, such as 'landless workers, small farmers, sharecroppers, pastoralists, fish folk and the peri-urban poor'.

According to Eric Holt-Giménez (2010), Via Campesina in a meeting in Spain made a declaration highly castigating international bodies like the IMF, The World Bank, WHO and CGIR directly responsible for food crisis and reaffirming the call for food sovereignty with imposition of an 'end to land grabs' aimed at the production of biofuels and foreign food production along with a call on international communities to reject the Green Revolution responsible for a tremendous loss of biodiversity along with authorising other evils like loss of local seeds. On the contrary, this gave overwhelming support for the 'seminal assessment' of the findings of the UN sponsored team of experts, the International Assessment of Agricultural Knowledge, Science and Technology for Development (IAAKSTD), represented by more than four hundred scientists and development experts from more than eighty countries, who concluded that: 'there is an urgent need to increase and strengthen further research and adoption of locally appropriate and democratically controlled agroecological methods of production relying on local expertise, local germ plasm, and farmer-manged local seed systems.'[15]

Henceforward, it came to be a historic achievement for the local farmers to act as the saviour of mankind as the supplier of food to eat and thus remove hunger and not to use it for industrial purposes (though still depending largely on decisions of different communities in power).

As reported by Eric Holt-Giménez (2010), one of the most successful examples of this movement for sustainable agriculture is provided by the farmers of Latin America made up of 'hundreds of peasant technicians' engaged in farming in over a dozen countries applying 'organic matter to soils and by implementing soil and water conservation techniques' they frequently obtained yield growths of 100-400 per cent.

> Initial methods of composting, soil and water conservation, and seed selection soon developed into a sophisticated '*basket* of sustainable technologies and agroecological management approaches that included green manures, crop diversification, integrated pest management, biological weed control, reforestation, and agrobiodiversity management at farm and watershed scales.'[16]

Such an all-comprehensive approach promised revolutionary changes leading to the spread of highly successful farmer-generated agroecological practices worldwide and achieved record-breaking results through the application of the System of Rice Intensification (SRI) developed in Madagascar raising the record-breaking yield of rice as high as '8 metric tonnes per hectare to spread to a million farmers in over two dozen countries'. It is further interesting to know that Wayne Roberts (2013, New Ed.) refers to the unique success achieved by a small holder rural farmer in Bihar (India) growing 22.4 tonnes of rice on one hectare of land following the principle of Madagascar.[17] This is, however, not the lone instance of deriving unchallengeable success for two other practising small holder farmers in the same state (Bihar, India) achieved a world-record in potato-fields and another farmer doing the same to wheat-yields through indigenous methods.

In this regard, we should refer again to Jules Pretty (2010) to review the situation, to find out how far the world depends on the small holder farmers for food-situation better for all, irrespective of rich and poor, without keeping the two worlds – North and South – apart. And it is in the developing countries that significant progress in sustainable agroecosystems have been made consisting of 'four different mechanisms':

(a) 'Home garden intensification with vegetables and/or tree crops, vegetables on rice field embarkments, and introduction of fishponds or a dairy cow;

(b) Addition of a new productive element to a farm system, such as fish or shrimp in paddy rice or agroforestry, which provides a boost to farm production and/or income, but which does not necessarily affect cereal productivity;

(c) Better use of nature to increase total farm production, especially water (by water harvesting and irrigation scheduling) and land (by reclamation of degraded land), leading to additional new dryland crops and/or increased supply of water for irrigated crops, and thus increasing cropping intensity; and

(d) Improvements in per hectare yields of staples through the introduction of new generative elements into farm systems such as legumes and integrated pest management, and new and locally appropriate crop varieties and animal feeds.'[18]

According to Jules Pretty (2010), such a management system will undoubtedly improve domestic consumption as well as improve local food sales without affecting the 'per hectare yields of cereals'. Besides, these sustainable agroecosystems contribute to the enhancement of local environments, strengthen communities, and develop human capacities with positive side effects, the most important of which is:

> improvement to the ecosystem including increased water retention in soils, improvement in water table (with more and cleaner drinking water in the dry season), reduced soil erosion combined with more organic matter in soils, leading to more carbon sequestration, healthier soils, greater productivity, and increased agrobiodiversity.[19]

But the success of the entire system – highly promising for the poorest – depends on the farmers themselves who should be capable of implementing it through 'social learning' depending on the farmers 'ecological literacy' as it benefits the poor as well as the environments in the developing countries.

VI. The Role of Agriculture

Speaking broadly, agriculture has special powers in reducing poverty. For instance in China agriculture has been more effective in reducing 3.5 times more poverty 'than growth outside agriculture and 2.7 times more in Latin America'. To make food available for all, including 800 million people still remaining food-insecure, cereal production, according to The World Bank Report, 2008, has to increase by 50 per cent and meat production by 85 per cent from 2000 to 2030 which requires improvement of assets 'of the rural poor consisting of smallholding farmers to make them more competitive and sustainable'. Such assets are land, water, education (training) and health. According to the same source (The World Bank, 2008), 60 per cent of the rural population in the developing world have good access to the global markets by diversifying to higher-value products in the perspective of dietary changes 'shifting away from cereals, roots, tubers and pulses to livestock products, vegetable oils, fruits and vegetables'. As a matter of fact, 'fruits and vegetables are one of the fastest growing agricultural markets in developing countries, with production increasing by 3.6 per cent a year for fruits and 5.5 per cent for vegetables over 1980-2004, 58 per cent of the increase coming from China alone, 38 per cent from all other developing countries, with 4 per cent contributed by the developed countries. In India, fruits and vegetables came to be the most important growth sector in crop production in the 1990s. Compared to cereals, the return from horticulture on land is 'about 10-fold' creating more off farm jobs in 'processing, packaging and marketing'.

Hand in hand such high value products like horticulture, livestock, fish, cut-flowers and organic products represent '47 per cent of all developing country exports, much more than traditional exports of coffee, tea and cotton comprising about 21 per cent.'[20]

Contributions of R&D to Agricultural Development with Great Benefit to Small holder Farmers:

> 'The technological challenges facing agriculture in the 21st Century are probably even more daunting than those in recent decades. With the increasing scarcity of land and water, productivity gains will be the main source of growth in agriculture and the primary means to satisfy increased demand for food and agricultural production. With globalisation and new supply-chains, farmers and countries need to continually innovate to respond to changing market demands and stay competitive.'[21]

This is obviously a timely warning for farmers everywhere unmistakably in the developing countries not only to stay alive to the changing situation affecting not only their only livelihood but also for the world as a whole to meet the growing demand for food.

Scientific plant breeding starting with improved varieties of seeds suited to small holders in subtropical and tropical areas beginning in the1960s made a revolutionary contribution to agricultural development in developing countries with successful introduction of more than 8,000 crop varieties over the past four decades leading to a higher yield growth exceeding that of Green Revolution amounting to 'as much as 50 per cent in the 1980s and 1990s'.

In India, for example, 80 per cent of the cereal area is planted with improved varieties of seeds only half of which is irrigated, and the improved varieties of wheat provide a 1 per cent increase in yield annually while globally the area has doubled since 1981 brought under improved varieties of seeds mostly in rain-fed areas. Even Sub-Saharan Africa as well shows promises in this respect with crops like maize, cassava, rice, and beans adopted by small holders with almost universal coverage for maize.

We need to be reminded that since food security and livelihood of all farmers are directly dependent on yield-stability, improved varieties make yields more stable providing resistance against vulnerability to pest, disease outbreaks, droughts, and other stresses. It is very important to know that the 'annual benefits from better yield stability in maize and wheat alone' come to about $300 million.

However, to maintain yield stability endangered by a constant fight against outbreaks from new races of pathogens constant vigilance is essential requiring investment in *maintenance research* making it obligatory to spend preferably the larger portion (at least a third to half) of the current investments in R&D in continuing its role in furthering productivity advances. The absolute necessity of R&D's hold can be realised from the success in tackling the recent 'outbreak of bacterial wilt in East Africa's Great Lakes Region with' 'a genetically engineered variety with resistance to the disease', so that the region boasts of the highest per capita consumption of bananas in the world.

As mentioned earlier, conservation or zero tillage has turned out to be the 'most dramatic technological revolutions' in crop management, in which eliminating tillage keeps crop residues as ground cover thus saving labour and energy, and conserves the soil and increases tolerance to drought, and reducing GHG emissions, though by making control of weeds, pests and diseases more complex, zero-tillage 'requires some provision of herbicides'.

However, the zero-tillage cultivation has been a more popular practice in Latin America, where it has been practised in more than '40 million ha or about 43 per cent of arable land' in Latin America, particularly in Argentina and Brazil, by large and midsize farmers to small farmers in

Southern Brazil, while 'an estimated 1,00,000 small holders' in Ghana have turned to this practice. As observed by The World Bank 2008 Report (Agriculture for Development) 'zero-tillage increases yield-rates of wheat, reducing production cost by 10 per cent, water-use by about 1 million litres per hectare, improves soil structure, fertility, and biological properties' as well as 'reduces the incidence of weeds and some other pests'. The report, quoting from another report (Malik, Yadav and Singh, 2005; Paris, 2003) mentions that zero-tillage practice is now a widely adopted one in the Indo-Gangetic Plain with about '0.8 million ha planted with rice and wheat estimated with a rate of return of 57 per cent on an investment of $3.5 million'.

Another resource-conserving technology is the introduction of legumes biologically fixing nitrogen which reduces needs for chemical fertilizers and results in yield-gain besides reducing soil erosion. As observed further in the report mentioned, the practices of 'rotational fallows and permanent intercropping have found to be highly beneficial as means of soil fertility management keeping African soils productive and healthy.'[22]

Most important of all is to keep R&D more responsive to farmers which requires farmers' direct participation in decision-making at R&D level. This helps in identifying 'constraints, poor indigenous knowledge, and aggregate technological demands.'

> 'Farmers' organisations have demonstrated strong interest in such partnerships. One approach empowers farmers by formally including them in governing councils of research organisations. This generally produces results only if the system is decentralised and farmers have a controlling interest in resource allocation – giving them the power to approve research projects and programmes as in Mexico.'[23]

In short, since agriculture, particularly engaged in providing for human-consumption rather than for industrial purposes like biofuel, or animal-feed, remains as the domain of small holder farmers in developing countries and should have 'direct access to the activities of the R&D in preparation as well as execution of policies without which feeding the burgeoning global population will not come to fruition.'

NOTES

***Food Agriculture Organisation (FAO) of the United Nations.** *The Future of Food Agriculture: Trends and Challenges*. Rome. 2017, pp. 112, 116.

'Food losses and waste also hold back the transition to environmentally sustainable food systems. They represent a considerable waste of land, water, energy, and agricultural inputs, and cause the emission of millions of tonnes of greenhouse gases. Future efforts to address climate change will need to find ways to reduce food losses and waste. Because food production is responsible for a large share of GHG emissions, reducing food losses and waste contributes to climate change mitigation. At the same time because climate change threatens food production in many food insecure areas, reducing food losses and waste can be an important part of climate change adaptation strategies .... In dealing with the problem of food waste, technological fixes do not offer lasting solutions. Responses must address the attitudes and actions of a range of stakeholders throughout the food supply chain. In high-income countries, food waste is caused mainly by consumer behaviour and economic decisions, and by policies and regulations related to other sectors. For example, agricultural studies may encourage the production of surplus food crops. This excess production helps contain prices but also causes less attention to be paid to food waste, both by value chain stakeholders and by consumers. Food waste is most often caused by retailers and consumers over-purchasing and then simply throwing away perfectly edible foodstuffs. In addition, food safety and quality standards may remove from the supply chain food that is still safe to eat. At the consumer level, inadequate planning of purchases and failure to use food before its expiry date also contribute to large waste.'

1. Food Agriculture Organisation (FAO) of the United Nations. *The Future of Food Agriculture: Trends and Challenges*. Rome, 2017, pp. 4-5.
2. *Ibid.*, pp. 32-34, 36, 38.
3. *Ibid.*, pp. 40-41.
4. *Ibid.*, p. 48.
5. *Ibid.*, p. 49.
6. *Ibid.*, p. 112.
7. Alexandratos, N., Bruinsma, J. *World Agriculture Towards 2030/2050.* FAO ESA Working Paper 12.3, 2012, p. 99.
8. *Ibid.*, p. 104.
9. *Ibid.*, p. 105.
10. *Ibid.*, pp. 105-109.
11. Holt-Giménez, Eric. "From Food Crisis to Food Sovereignty: The Challenge of Social Movements." in *Agriculture and Food Crisis: Conflict, Resistance, and Renewal*, ed. F. Magdoff, B. Tokar. New York: Monthly Review Press, 2010, p. 211.
12. Ghosh, M. Economics Department, Visva-Bharati University, West Bengal. *The Statesman*. Kolkata. 10 April, 2018.
13. Alexandratos, N, Bruinsma, J. *World Agriculture Towards 2030/2050.* FAO ESA Working Paper 12.3, 2012, pp. 78-79.
14. Holt-Giménez. Eric. "From Food Crisis to Food Sovereignty: The Challenge of Social Movements." in *Agriculture and Food Crisis: Conflict,*

Resistance, and Renewal, ed. F. Magdoff, B. Tokar. New York: Monthly Review Press, 2010, pp. 211-213

15. *Ibid.*, p. 215.
16. *Ibid.*, p. 216.
17. Roberts, Wayne. *The No-Nonsense Guide to World Food.* U.K.: New Internationalist, 2013, p. 143.
18. Pretty, Jules. "Can Ecological Agriculture Feed Nine Billion People?" in in *Agriculture and Food Crisis: Conflict, Resistance, and Renewal*, ed. F. Magdoff, B. Tokar. New York: Monthly Review Press, 2010, pp. 293-294.
19. *Ibid.*, p. 295.
20. Agriculture for Development. *The World Bank: Development Report 2008*, pp. 6-7, 58, 60.
21. *Ibid.*, p. 158.
22. *Ibid.*, pp. 160-161 and 164-165.
23. *Ibid.*, p. 172.

7
Conclusion

In drawing a conclusion on agricultural development and its role in feeding people on earth leaving no one hungry by mid-twenty first century, one surely has to look back to have a proper idea about its development during the long period of its evolution and progress through a period of about ten thousand years or so since it made its sporadic appearance in certain locations. As a matter of fact, the appearance of agricultural practices is one of the greatest, if not the greatest of all historical events in human history creating a completely new path to follow hence forward. We should call it 'accident' since it was not a premeditated action but happened accidentally more as a part of natural process than as a manifestation of human efforts. There is, however, no record in support of the evolutionary process and we have to imagine how it came about. In analysing the situation leading to the emergence of its progress we have to believe that nature's process of regeneration struck people in those days watching almost unaware that seeds of plants falling on the ground sprouted after an interval into life again taking the exact form of the parental plant from which the seed had fallen after the fruit had matured and fallen seeds on the ground took rebirth into the same plant revealing the process of life sustained through generations.In this regard, it is believed that it was women of the earlier generation who had time enough to watch and discover the natural process of rebirth in plant life when men were too busy in foraging for food for survival with no time in hand to watch what was happening areound.[1]

This gives a background of the development of the practice of agriculture some ten thousand years or so ago which made one of the most revolutionary changes in human life, for the first time making human life sedentary instead of a constant movement following the supply of food. This is the first-time man could settle down permanently at one place producing food for his own survival and thus presumably proliferate in numbers without having to be exposed to the imminent danger of life pursuing food and become victims of the pursued.

Though, in all probability, the practice of growing food, developed, may be simultaneously or rather in around a period over a few centuries (people borrowing ideas when located not too far from each other).Its original development is believed to have centred around a region known as 'Levant' today* lying more or less in a similar climatic belt providing a natural setting to accede to human needs of similar nature. Modern Syria and Lebanon in the Middle East represent the Levant, which in all probability may be taken as the cradle of the origin and sustainability as well as the processor of agriculture following trade routes. There is archaeological evidence indicating the Middle East as the cradle of earliest agricultural practices (mentioned in the 'Introduction' of this book), having been the homeland of wheat and barley besides many other crops where the plough was first invented. It is highly relevant to know that agricultural implements dating back from 9500 B.C. have been discovered at Jericho in the Jordan Valley, showing further possibilities of agricultural practices spreading from Asia Minor to Western Iran, Turkey, Cyprus and into Egypt a little before 4500 B.C. while in all probability plant domestication spread to Europe 'prior to 6000 B.C. with the introduction of einkorn and barley as the major crops'. On the other hand, agricultural practices spread from Iraq and Iran to Southern Asia in about 3500 B.C. and by 2500 B.C. crops like bread wheat, barley, dates and perhaps rice were grown in the flourishing civilisation of the Indus Valley. In the lack of any authentic supportive evidence (after this) about the nature of agricultural progress, it is not possible to know what sort of changes took place in the practice of agriculture until we arrive in the modern period when agriculture came to be a dominant practice of the majority of people making it their livelihood and feeding the growing population the worldover. It, however, did not take long for man to misuse natural limited resources, like soil, water and land on which agriculture thrived, and this happened at a faster rate in the modern period with ever-increasing demand for more comfort in life with reckless handling of its own limit. According to a study called, 'Limits of Growth' by 'The Club of Rome'[2], there are about '3.2 billion hectares of potentially suitable land for agriculture' about half of which – the richest, most accessible is under cultivation, while the remaining land needed high capital inputs to make that productive.

The same 'Report' mentioned that each person requires 0.08 ha per head to cover all requirements including food, housing, roads, waste disposal, power lines and other essential uses. Since the overall food production has to increase following growth of global population 'advances in agricultural techniques in future' are likely to increase yield rates of food crops twice or even four times with large investments in 'tractors, fertilizers and irrigation systems'. Hence, the supply of more food in the opinion of experts (The Club of Rome) was dependent on 'land and fresh water and agricultural capital'. According to Dieter Helm (2015)[3], in terms of feeding 9 or 10 billion people in around mid-twenty first century would require revolutionary changes in agricultural practices keeping in mind the growing changes in the nature or contents of human-consumption with double of the requirement of meat, fish and nutritional food spurred by further rising income in many of the developing countries, while the situation is further aggravated by increasing proportion of food crops like corn (maize), sugarcane, palm oil and even wheat (in Britain) converted to the production of ethanol (food for cars including 40 per cent of maize in the US). Apart from affecting carbon emission with further rise, all these efforts through agricultural expansion are mostly at the cost of tropical/subtropical forests so long acting as large carbon reservoirs and will destroy the 25 biodiversity hotspots of the world originally covering about 12 per cent of the global land surface reduced henceforward to 1.4 per cent along with the loss of primary vegetation besides severe decline of animal life amounting already to about 52 per cent of their total since 1970.

Dieter Helm (2015) points out that if we get a change in food habits with greater stress on plant food (vegetarian diets) it will cut down consumption of food crops as animal-feed along with lessening carbon emissions effectively thereby making a reduction in the rise of temperature enhanced further with the adoption of compulsory recycling of water and materials along with compulsory introduction of drip-irrigation, no-till farming practices and drastic reduction in the use of chemical fertilizers and harmful insecticides with highly positive impacts on making agricultural practices more amenable to man's role as food-producer. Such an attempt may be considered as the most important change in the character of agricultural practices giving it a more human face than when it was turning into serving the purposes of giant corporate houses trying to convert it into industrial uses for their own benefit, keeping growing number of people hungry as well as malnourished.

As a remedy to the situation, improvement of agricultural productivity came to be of prime concern. Such an approach, based entirely on science and technology, called a 'biological process' starting in the 1960s succeeded in developing improved varieties of animal breeds adapted to sub-tropical and tropical areas made high returns with pro-poor impacts with the promise of better breeds in future. During the same time 'scientific plant breeding developed improved varieties of seeds suited to small holder and sub-tropical and tropical areas contributing semi-dwarf varieties of rice and wheat and improved varieties of maize and contributed as much as 50 per cent yield-growth without which global food prices would have been 18-21 per cent higher in 2000 making '13-15 million children undernourished'. The genetic improvement suitable for both irrigated and rain-fed crops has continued to contribute to yield-gains with 80 per cent of cereal area in India half of which is irrigated planted with these genetically improved crops while globally the area planted with such seeds more than doubled since 1981, increasing yield by 1 per cent in rain-fed areas. Even Sub-Saharan Africa is deriving great benefit of this biological breeding through R&D for some major crops like improved varieties of disease-resistant strains of cassava, rice and beans combining the high-yield potential of Asian rice with resistance to 'weeds, pests and diseases and water-stress' increasing yield, now cultivated on 2,00,000 ha in Africa while new varieties of beans with multiple stress resistance are grown widely in eastern, central and southern Africa by about 10 million farmers, mostly women.

Thus, the major emphasis laid by R&D was on improving the management of crops, livestock and on reducing clash with the natural resource system much of which was directed to soil and water management along with agroecological approaches stressing reduction in the uses of agricultural chemicals. In order to achieve these goals, the methods pursued included conservation, or zero-tillage, maintaining crop residues as ground cover, the practice itself giving rise to many other advantages like saving labour and energy, conserving soil, increasing drought-tolerance, and reducing greenhouse gas emissions. Importance of this system is shown by the fact that in Latin America, Brazil and Argentina together use zero-tillage on more than 40 million ha, or 43 per cent of the arable land which is also practised by smallholding farmers in Southern Brazil, in Ghana, and is going to be rapidly adopted by small farmers in the Indo-Gangetic Plains adopting various other parts of the technology as well including rotation, seeds, chemicals and machinery. Another important step taken is biologically fixing nitrogen in the soil with the introduction of legumes while in South Africa fast-growing 'fertilizer trees' not only improve soil-quality but also check soil erosion. At the same time, improved methods of pest management with reduction in the use of dangerous pesticides along with the use of legumes replacing fertilizers coupled with 'rotational fallow' have more than doubled maize-yield in Southern Africa. The practice of rotational fallows and permanent intercropping have benefited 12 million smallholder farmers in eastern and southern Africa protecting soil fertility and keeping it productive and healthy.

However, the major handicap in improving agricultural practices is a great reluctance on the part of governments or men in power in investing in R&D which stalls more progress in their quest for agricultural improvement and thus acts as the greatest impediment towards production of food to save man from hunger. In this respect, it is highly praiseworthy to note that some of the developing countries like China, India and to a less extent Brazil have rapidly increased their spending on R&D over the past two decades while Sub-Saharan Africa has lacked initiative in this although benefiting from the spill-over from other developing countries, though the major crops she depends on being cassava, yams, millets, plantain and teff do not match with those in other developing countries.

It is, however, very important to note that the smallholder farming system, the prevailing agricultural system in developing countries providing food for their own and others' consumption suffers from lack of access, inputs, credit, and other ownership problems along with bearing risks of climate instability. In order to tackle these challenges, the smallholder farmers are encouraged to stress demand-led agricultural extension taking help from agricultural research besides exchanging information on seeds on the basis of farmer-to-farmer networks. In other words, such a development replicates the La Campes in a approach as the basis for the Food Sovereignty movement.

All these exercises are finally aimed at making food available for everybody, including the smallholder rural farmers in the developing countries who grow it for their own need as well as providing for others who need it at reasonable prices. The entire exercise beginning from production to sale acts as a process giving him/her freedom from hunger.

The most important question that arises here is whether the poor in rural areas specifically in the developing countries making a living based on agricultural activities have any prospect of facing the great challenge of feeding the growing population in future. Jules Pretty (2010)[4] feels quite optimistic that it is possible provided future steps are taken in the right direction. In his opinion, there is enough staple food produced in the world but as long as vitally wrong policies are involved in their consumption, such as 37 per cent of it at present being fed to animals in the developing countries while 73 per cent in the industrialised countries besides a substantial part of it being wasted by the rich during consumption, there will always remain a food scarcity. In this respect, a new approach to the productive process based on locally available and/or cheap technologies combined with 'making the best of natural, social and human resources' using 'best available technologies along with best ecological management may come to our rescue'. A major part of this new approach involves using nitrogen-fixing legumes replacing fertilizers and putting emphasis on natural enemies instead of using pesticides and reduced tillage or no tillage 'besides making better use of land, water and biodiversity and technologies'.

One of the most important identifiable instance of smallholders farming is its feminization, more attached to agricultural activities even when the male partner may have gone out to join some other profession.

'In many countries in the Near East and North Africa, Central Asia, South Asia and Latin America, the female share of agricultural employment has increased significantly in recent decades, and women have become the majority of those employed in the sector.... The female share of agricultural employment is also rising in South and Central Asia'[6] (Table Con.1).

Table Con.1 gives us a comprehensive idea about feminisation of agriculture of the developing countries showing that out of 47 selected countries, 26 in Africa, 17 in Asia and 4 in Latin America, the percentage of females was higher between 1980 and 2010 in 14 of them while in 15 of them the share further increased in 2010. However, Lesotho (in Africa) having the highest of female share throughout the period (1980-2010) having 67.3 to 72.0 per cent of the total labour force records a slowly declining share while Palestine in Western Asia claims the highest share among all Asiatic countries ranging from 64.1 to 72.5 per cent in 2010 (higher than Lesotho) though, in general, West Asian countries have higher feminine shares than other Asian countries in sharp contrast to Latin America where the share is very low in general.

Table Con. 1: Female Share of Economically Active Population in Agriculture in 1980, 1995 and 2010 (%)

	1980	1995	2010
	AFRICA		
Burundi	55.9	55.9	**56.0**
Comoros	50.0	50.3	**52.0**
Madagascar	54.7	53.9	53.5
Malawi	56.7	56.1	**59.2**
Mozambique	58.6	63.4	**65.2**
Rwanda	55.3	56.1	**57.0**
United Republic of Tanzania	53.7	54.1	**55.0**

contd...

	1980	1995	2010
Zimbabwe	54.3	55.3	53.3
Angola	52.4	52.6	**55.0**
Chad	28.9	50.8	**56.9**
Congo	56.6	60.0	56.5
Botswana	46.6	52.4	**56.9**
Lesotho	72.0	68.2	67.3
South Africa	37.1	31.1	29.6
Swaziland	58.5	60.7	54.3
Benin	34.5	41.1	39.6
Gambia	50.6	51.2	**53.3**
Mauritania	47.6	49.2	**53.9**
Senegal	44.9	45.5	**47.4**
Sierra Leone	59.0	58.5	**61.7**
Algeria	41.5	50.4	53.7
Libya	37.2	50.0	**69.9**
Morocco	29.0	38.9	**47.7**
Sudan	32.5	34.9	**39.5**
Tunisia	27.1	34.4	32.8
Western Sahara	41.1	47.8	**53.7**
ASIA			
Cambodia	57.3	54.9	51.2
Indonesia	33.7	39.9	**39.3**
Lao Peoples' Democratic Republic	51.3	51.8	**52.3**
Vietnam	50.7	51.0	49.1
Bangladesh	42.4	44.5	**51.0**
Bhutan	26.3	19.4	**34.7**
Iran (Islamic Republic of)	25.2	33.9	**46.4**
Nepal	35.4	42.2	**48.1**
Papua New Guinea	47.9	53.5	**55.8**
Pakistan	12.2	18.4	**29.6**

contd...

	1980	1995	2010
WESTERN ASIA			
Azerbaijan	-	53.8	**53.9**
Iraq	29.7	38.2	**59.3**
Jordan	41.9	44.3	**62.2**
Palestine	64.8	64.1	**72.5**
Syrian Arab Republic	31.7	50.7	**60.7**
Turkey	40.4	48.2	**52.3**
Yemen	29.3	34.4	**40.1**
LATIN AMERICA			
Chile	9.2	10.6	**14.2**
Colombia	19.5	19.9	**24.8**
Ecuador	14.0	17.6	**24.8**
Peru	19.0	27.0	**31.3**

Source: Food Agriculture Organisation (FAO) of the United Nations. *The Future of Food Agriculture: Trends and Challenges*. Rome. 2017. Table 11.2. p. 103.
Note: Increase in 2010 highlighted in bold.

In other words, farming as a livelihood is a family job, both husband and wife, or any other female member of the family working together in the fields even when the male member has moved on, usually for a short period to some other source of earning and thus putting the entire responsibility on the female members of the family. Growing crops for feeding themselves and marketing the surplus, if there was any, came to be a way of life for these smallholder farmers, the practice for the overwhelming majority of rural people in the developing world, accepted as the only way of life giving no choice of change since any other option is beyond the capability of the small family-units to tackle. But they exhibit a remarkable degree of resilience for survival as proved by remaining stable over the last forty years (1970 to 2010, no data being available afterwards) steady in supplying food for the Global South, including '90 per cent of all the food requirements of Africa' (Eric Holt-Giménez, 2010), as mentioned earlier.

What is, however, the important relevant issue about their present inescapable position of getting stuck in food-production as a permanent way of living goes back to the most shocking event of their origin as a product of European Colonisation when a large section of the rural population all-over the Indian Subcontinent were engaged in handicraft activities producing inimitable products particularly of cotton and different metals enjoying a global market for their extremely high quality. The British having colonised India made a determined policy of destroying the cottage industry and did it with all cruelty even going to the extent of cutting off the thumb of the master craftsmen so that they could no longer use the loom and thus turning a large section of rural population completely bereft of any means of living so they took up tilling soils as the only way of survival. Thus, European Colonisation to make way for their own products of much inferior quality replacing the native ones of extraordinary high quality destroyed for good, turning this section engaged till then in making a very strong base for the national economy not only made them lose their own mooring but also dragged down the country's entire economy to a very low-level following policies as determined by the colonisers which were nothing short of a relentless loot and pillage. An idea about the nature of British rule in India during colonial era may be obtained from the fact, mentioned earlier, that there had been 31 serious famines in 120 years of British rule against only17 recorded famines in the entire previous two millennia (Mike Davis, 2014 Reprint). That the Indian economy had been much stronger is revealed by the fact that 'South Indian labourers had higher earnings than their British counterparts in the eighteenth century and lived in greater financial security' is enough proof to justify the claim that India's economy was on a stronger footing even then when colonisation was in process (Mike Davis, 2014 Reprint). This is further supported by the share of World GDP as shown in the Table Con.2.

Table Con. 2: Share of World GDP

	1700	**1820**	**1890**	**1952**
China	23.1	32.4	13.2	5.2
India	22.6	15.7	11.0	3.8
Europe	23.3	26.6	40.3	29.7

Source: Davis, Mike. *Late Victorian Holocausts: El Nino Famines and the Making of the Third World.* Version, London and New York. 2014 (Reprint). p. 293. Taken from Madison, Angus, *Chinese Economic Performance in the Long Run.* Paris. 1998, p. 40.

The slow economic downward trend is shown in Table Con. 2 for both China and India as a result of colonisation for India and economic blockade by colonising powers for China.

In reality European colonisation represents one of the darkest episodes of human history changing the life of almost half of humanity pushing it back with matchless cunning, cruelty and brutality into perpetual 'have nots'. In spite of the withdrawal of subjugation mostly during the post-Second World War period the limitless, unending economic plunder of these countries means they have been left nothing as economic capital for building up their economy. Almost half of humanity having been treated as a source of strengthening the 'Home Country' of the colonisers, they also saw to it that even when free of colonial rule the colonies remained economically bankrupt. India provides a very strong example of this. It is very interesting to know that the railways introduced in India as early as 1853 even when they were not built in many European countries were used to carry raw materials from India to Britain for the factories and bring back low-quality British products for the Indian market. The irrigation projects installed in the country were utilised for the production of crops like cotton, required by the Home Country getting cheap raw materials to be used by their factories at Home and sold in the Indian market. The British resisted the development of science and technology** and instead encouraged the production of a so-called 'Babu Class' represented by people with a low level of education to serve their administrative set-up.

As a result of all this almost half of humanity colonised by European colonisers remained economically and even technically crippled to stand up to full stature when they were left to themselves since it was not only economic but also social crippling very much cunningly done, vitiating the situation and very consciously utilising the difference in race, religion, customs, and language.

The smallholder farmers, or landless peasants etc. permanently attached to agriculture as their inseparable livelihood drifted away from their original occupation of rural-crafts sustaining rural economy with their extremely high-quality products enjoying a global market lost the mooring once and for all when the colonisers intentionally destroyed this structure of rural handicrafts and these people having no alternative drifted to farming as the family-profession on a permanent basis.

NOTES

*__Encyclopaedia Americana.__ Volume 17. p. 260.
Levant: The lands bordering the eastern Mediterranean. The French word for 'rising' or 'sunrise', it is still sometimes applied to Syria and Lebanon, which were French mandates known as Levant States.

**__Geddes, Patrick.__ *The Life and Work of Sir Jagadish Chandra Bose*. Longmans, Green and Co., London. 1920.
Sir J.C. Bose (Jagadish Chandra Bose) was born on November 30, 1858 in Bikrampur, Dacca district (now in Bangladesh) was one of the leading scientists, at least a hundred years ahead of his contemporaries in the world,

ignored for Nobel Prize twice, once in 1909 when the discovery on Electro-Physiology discovered earlier by him in 1907 went to Gugliemo Marconi succeeding later than Bose* while another most epoch-making discovery of life in plant (demonstrated many times), as referred to by Patrick Geddes in his biography *The Life and Work of Sir Jagadish Chandra Bose*. Longmans, Green and Co., London. 1920. Sir J.C. Bose died at Giridih on 23 November, 1937.

*This is corroborated by Marconi's grandson, Francesco Paresce Marconi while visiting J.C. Bose Institute and Museum in Kolkata on 30 November, 2006, in the following words: 'This is a wonderful experience for me since Sir J.C. Bose actually preformed the first wireless experiment before that of my grandfather Gugliemo Marconi'.

1. Read for more information: Hawkes, J., Woolley, L. *History of Mankind: Cultural and Scientific Development. Prehistory and the Beginning of Civilization*. London: George Allen and Unwin Ltd., 1963.
2. Meadows, D.H., Meadows, D.L., Panders, J., Behren, III, W.W. *The Limits to Growth: A Report for The Club of Rome's Project on the Predicament of Mankind*. New York: Universe Books, 1972.
3. Helm, Dieter. *Natural Capital: Valuing the Planet*. New Haven and London: Yale University Press, 2015.
4. Pretty, Jules. "Can Ecological Agriculture Feed Nine Billion People?" in in *Agriculture and Food Crisis: Conflict, Resistance, and Renewal*, ed. F. Magdoff, B. Tokar. New York: Monthly Review Press, 2010.
5. Food Agriculture Organisation (FAO) of the United Nations. *The Future of Food Agriculture: Trends and Challenges*. Rome. 2017, pp. 103-104.

Bibliography

Agriculture for Development. The World Bank: World Development Report, 2008.

Alexandratos, N. and Brunisma, J. *World Agriculture towards 2030/2050*. ESA Working Paper 12(03), Rome, 2012.

Banerjee, A.V. and Duflo, E. *Poor Economics: Rethinking Poverty & Ways to End It*. India: Penguin Books, 2013.

Bhattacharya, Bimalendu. *Geography of Deprivation: An Unfair World*. New Delhi: Concept Publishing House Ltd., 2014.

Bhattchharya, Bimalendu. *Food: The Growing Challenge for Human Survival*. New Delhi: Concept Publishing Company Pvt. Ltd., 2020.

Brown, Lester R. *Eco-Economy: Building an Economy for the Earth*. Hyderabad: Orient Longman, 2004.

Brown, Lester R. *Full Planet, Empty Plates*. New York, London: W.W. Norton & Company, 2012.

Clark, Wilson. *Energy for Survival: The Alternative is Extinction*. 1973. Bombay: Allied Publishers Limited, 1977.

Davis, M. *Late Victorian Holocausts: El Nino Famines and the Making of the Third World*. London and New York: Verso, 2014.

Dyson, Tim. *Population and Food: Global Trends and Future Prospects*. London: Routledge, 1996.

Drewnowski Adam, Popkin Barry M. "The Nutrition Transition: New Trends in the Global Diet." *Nutrition Reviews*, 55.2 (1997), pp. 31-43.

Ehrlick, P.R., Ehrlick, A.H., Daly, G.C. "Food security, population, and environment." *Population and Development Review*, 19 (1993), pp. 1-32.

Encyclopaedia Americana, New York: Americana Corporation, Vol. 1, 1973.

Fitzpatrick, Tony. *Climate Change and Poverty: A New Agenda for 'Developed Nations'*. Bristol: Policy Press, 2014.

Foley, J.A., Ramamurthi, N., Brauman, K.A., Cassidy, E.S., Gerber, J.S., Johnston, M., Mueller, N.D., O'Connell, C., Ray, D.K., West, P.C., Balzer, C., Bennett, E.M., Carpenter, S.R., Hill, J., Moufreda, C., Polasky, S., Rockstrõm, J., Sheehan, J., Tilman, D., Zachs, D.P.M. "Solutions for a Cultivated Planet." *Nature,* 478 (2011), pp. 337-342.

Ghosh, M. Economics Department, Visva-Bharati University, West Bengal. *The Statesman*. Kolkata. 10 April, 2018.

Godfray, H., Charles, J., Beddington, J., Crute, I.R., Hadad, L., Lawrence, D., Muire, J. F., Pretty, J., Robinson, S., Thomas, S. M., Toulmin, C. "Food Security: The Challenge of Feeding 9 Billion People." *Science: New Series* 327. 5967 (2010), pp. 812-818.

Hawkes, J., Woolley, L. *History of Mankind: Cultural and Scientific Development. Prehistory and the Beginning of Civilization*. London: George Allen and Unwin Ltd., 1963.

Helm, Dieter. *Natural Capital: Valuing the Planet*. New Haven and London: Yale University Press, 2015.

Human Development Report 2007/2008. Fighting Climate Change: Human Solidarity in a Divided World. United Nations Development Programme.

Juma, Calestous. *The New Harvest: Agricultural Innovation in Africa*. Oxford: Oxford University Press, 2011.

Kearny, John. "Food Consumption Trends and Drivers." *Philosophical Transactions of the Royal Society Biological Sciences* 365.1554 (2010), pp. 2793-2807.

Licker, R., Johnston, M., Foley, J.A., Barford, C., Kucharik, C.J., Monfreda, C. and Ramamkutty, N. "Mind the Gap: how do climate and agricultural management explain the yield gap of grasslands around the world?" *Global Ecology and Biogeography* 19.6 (2010), pp. 769-782.

Magdoff, F. and B. Tokar, ed. *Agriculture and Food Crisis: Conflict, Resistance, and Renewal*. New York: Monthly Review Press, 2010.

Majumdar, R.C., Ray Chaudhuri, H. C., Datta, K. *An Advanced History of India*. Delhi: Macmillan, 1970.

Mazhar, F., Buckles, D., Satheesh, P.V., Akhtar., F. *Food Sovereignty and Uncultivated Biodiversity in South Asia: Essays on the Poverty of Food Policy and the Wealth of the Social Landscape*. New Delhi: Academic Foundation, 2007.

McCully, Patrick. *Silenced Rivers: The Ecology and Politics of Large Dams*. Hyderabad: Orient Longman, 1998.

Meadows, D.H., Meadows, D.L., Panders, J., Behren, III, W.W. *The Limits to Growth: A Report for The Club of Rome's Project on the Predicament of Mankind*. New York: Universe Books, 1972.

Monbiot, George. "We can go on eating like this." *The Guardian* U.K., 11 December, 2017.

People and Ecosystems: The Fraying Web of Life. United Nations Development Programme. United Nations World Resources 2000-2001.

Popkin, Barry M. "Relationship Between Shifts in Food System Dynamics and Acceleration of the Global Nutrition Transition." *Nutrition Reviews* 75.2 (2017), pp. 73-82.

Rhoads, Robert E. "The World's Food Supply at Risk." *The National Geographic Society* 179 (1991), pp. 74-107.

Roberts, Wayne. *The No-Nonsense Guide to World Food*. U.K.: New Internationalist, 2013.

Smil, Vaclav. *Feeding the World: A Challenge for the Twenty First Century*. Cambridge, Massachusetts: The MIT Press, 2002.

Stern, Nicholas. *A Blueprint for a Safer Planet: How to Manage Climate Change and Create a new Era of Progress and Prosperity*. London: The Bodley Head, 2009.

Sustainable Development in a Dynamic World, Transforming Institutions, Growth, and Quality of Life. *The World Bank: World Development Report 2003*.

The Future of Food Agriculture: Trends and Challenges. *Food Agriculture Organisation of the United Nations*, Rome, 2017.

Tilman, D., Balzer, C., Hill, J., Befort, B.L. "Global Food Demand and the Sustainable Intensification of Agriculture." *Proceeding of the National Academy of Sciences of the United States of America,* 108.50 (2011), pp. 20260-20264.

Tilman, D., Clark, M. "Global Diets Link Environmental Sustainability and Human Health." *Nature,* 515.7528 (2014), pp. 518-522.

Tilman, D., Clark, M. "Food, Agriculture and Environment: Can we feed to the world and save the Earth?" *The American Academy of Arts and Science,* 144.4 (2015), pp. 8-23.

Water for People Water for Life. *The United Nations World Water Development Report*, 2003.

Wittman, H., A. Demaris, N. Wiebe, ed. *Food Sovereignty: Reconnecting Food, Nature and Community*. U.K.: Pambazuka Press, 2011.

Index